Simplifying 3D Printing with OpenSCAD

Design, build, and test OpenSCAD programs to bring your ideas to life using 3D printers

Colin Dow

BIRMINGHAM—MUMBAI

Simplifying 3D Printing with OpenSCAD

Group Product Manager: Rohit Rajkumar
Publishing Product Manager: Aaron Tanna
Senior Editor: Hayden Edwards
Content Development Editor: Rashi Dubey
Technical Editor: Shubham Sharma
Copy Editor: Safis Editing
Project Coordinator: Rashika Ba
Proofreader: Safis Editing
Indexer: Sejal Dsilva
Production Designer: Shyam Sundar Korumilli
Marketing Coordinator: Elizabeth Varghese

First published: March 2022

Production reference: 1240222

Published by Packt Publishing Ltd.
Livery Place
35 Livery Street
Birmingham
B3 2PB, UK.

ISBN 978-1-80181-317-4

www.packt.com

This book is dedicated to my wife, Constance, and our sons, Maximillian and Jackson. This book would not be possible without their continuous love and support.

– Colin Dow

Contributors

About the author

Colin Dow has been 3D printing since 2013, starting with the laser-cut wooden frame version of the Ultimaker 3D printer. He has gone through a dozen or so 3D printers over the years, from MakerBots, PrintrBots, early Prusa i3s, Delta printers, and liquid resin printers. Colin has been working with OpenSCAD since 2014, using it with 3D printers to design and manufacture model rocketry parts for his model rocketry business. Through his aerospace workshops, he has introduced many students to 3D printing, including in-class demonstrations of 3D printing. Over the last few years, Colin has been designing and building automated drones for his drone start-up using 3D printers and OpenSCAD.

I would like to thank all the engineers, technicians, and innovators that have come before me. To paraphrase the great Sir Isaac Newton, "If I have seen further it is by standing on the shoulders of technology giants."

About the reviewers

Basil Dimakarakos is a self-employed IT professional acting as a support specialist with over a decade of experience in IT analysis of system needs, infrastructure design, and layout, with a Diploma of Computer and Network Engineering Technology.

His work as an IT professional has involved many roles: instructor for Office applications, analyzing business needs, configuring integration with peripherals, the design and layout of office networking systems, installing and configuring telecommunication tools, managing website/database access.

Basil has worked for IBM as a help-desk representative, TD bank as an internal user support assistant, Reed Exhibitions as an IT manager and support liaison, as an IT technician for SME Canada, and as a self-employed IT technician for several small to medium-size businesses over the years.

After work, he engages in a variety of activities including music recording, reading both fiction and non-fiction books, fixing old audio electronic gear, cycling, and cooking.

Most of all, he spends time exploring different technical and science-related media for new discoveries, inventions, gadgets, and ideas encompassing everything from artificial intelligence to zoology in the public media.

Constance Dow has been involved in technology for over 20 years. She started out in the software industry specializing in quality assurance and database configuration. She holds a BA from the University of Toronto as well as a computer science diploma from Sheridan College. Currently, she works for a large pharmaceutical company ensuring compliance with local regulations.

Gabriel Frampton is a modular origami designer and geometric artist with a parallel career as a basic science research assistant at the University of Texas at Austin Dell Medical School. He had previously been writing 3D geometry calculations by hand to develop new modular origami designs and discovered that the same principles could be applied to 3D printing by writing code using OpenSCAD. His work in 3D printing and geometric origami can be found by searching for `@foldedcrystals`.

Mats Tage Axelsson is a long-term Linux fanatic who has been designing his own CAD models since the last century. He has also worked for big telecoms companies, making your mobile phone systems bigger and better.

His most common activity nowadays is writing for Linux magazines on topics such as how to use and optimize your computer for your favorite applications.

Table of Contents

3
Printing Our First Object

Part 2: Learning OpenSCAD

4
Getting Started with OpenSCAD

5
Using Advanced Operations of OpenSCAD

6
Exploring Common OpenSCAD Libraries

Part 3: Projects

7

Creating a 3D-Printed Name Badge

8

Designing and Printing a Laptop Stand

9

Designing and Printing a Model Rocket

Part 4: The Future

10
The Future of 3D Printing and Design

Index

Other Books You May Enjoy

Preface

OpenSCAD is an open-source 3D design platform that helps you bring your designs to life. This book will show you how to make the best use of OpenSCAD to design and build objects using 3D printers.

This OpenSCAD book starts by taking you through the 3D printing technology, the software used for designing your objects, and an analysis of the G-code produced by the 3D printer slicer software. Complete with step-by-step explanations of essential concepts and real-world examples such as designing and printing a 3D name badge, model rocket, and laptop stand, the book helps you learn about 3D printers and how to set up a printing job. You'll design your objects using the OpenSCAD program that provides a robust and free 3D compiler at your fingertips. As you set up a 3D printer for a print job, you'll gain a solid understanding of how to configure the parameters to build well-defined designs.

By the end of this 3D printing book, you'll be ready to start designing and printing your own 3D printed products using OpenSCAD.

Who this book is for

This book is for engineers, hobbyists, teachers, 3D printing enthusiasts, and individuals working in the field of 3D printing. Basic knowledge of setting up and running 3D printers is assumed.

What this book covers

Chapter 1, Getting Started with 3D Printing, starts our exploration of 3D printers by looking at the Creality Ender 3 V2. We investigate the various parts that make up a 3D printer. We end the chapter with a look at the various materials that we can 3D print with.

Chapter 2, What Are Slicer Programs?, investigates G-code and slicer programs. We will control a 3D printer through the use of G-code before learning about slicing programs that turn 3D shapes into G-code for our 3D printer.

Chapter 3, Printing Our First Object, uses a 3D printer to print out objects. Knowledge gained from this chapter will be useful throughout the rest of the book as we bring our 3D ideas to life.

Chapter 4, *Getting Started with OpenSCAD*, explains how to create 3D shapes using OpenSCAD. We will compare OpenSCAD to other 3D design programs before we design a hook for a PVC pipe.

Chapter 5, *Using Advanced Operations of OpenSCAD*, continues to explore OpenSCAD as we learn ways to convert 2D shapes into 3D objects. We will take what we've learned and use it to design a Thumbs Up award trophy.

Chapter 6, *Exploring Common OpenSCAD Libraries*, looks into common libraries that we may use with OpenSCAD. We will use the knowledge gained to create a desk drawer that we can install under a table or desk.

Chapter 7, *Creating a 3D Printed Name Badge*, shows how to bend text around a circle. We will use this knowledge to create a name badge for a shop or conference. This is the first chapter in which we 3D print a design of our own.

Chapter 8, *Designing and Printing a Laptop Stand*, looks at designing a shape in Inkscape and importing it into OpenSCAD where we will turn it into a 3D shape. The project for this chapter is a laptop riser stand. We will design all the parts needed for the laptop riser stand, 3D-print them, and then assemble the stand using standard construction techniques.

Chapter 9, *Building a 3D Printed Model Rocket Using a Common Paper Tube*, takes a discarded paper towel tube and turns it into a model rocket. We will design and print out the motor mount, nose cone, and fins. Our model rocket will work with standard model rocket motors. This design is the first where we take measurements of other objects (the paper towel tube) and design parts around them.

Chapter 10, *The Future of 3D Printers and Design*, explores what the future may hold for 3D printers in the field of 3D printed homes and mass customization. For the final project of the book, we will design and print out a birdhouse.

To get the most out of this book

To get the most from this book, a background in the use of various software programs is desirable. In this book, we will be switching between various programs, such as OpenSCAD, Cura, and Inkscape, as we bring our 3D designs to life.

Software/hardware covered in the book	Operating system requirements
OpenSCAD	Windows, macOS, or Linux
Cura	Windows, macOS, or Linux
ideaMaker	Windows, macOS, or Linux
Late-model 3D printer. The Creality Ender 3 V2 will be used for demonstrations.	
3mm drill tap used to connect plates together in *Chapter 8*, a model rocket base plate in *Chapter 9*, and the bottom tray of a birdhouse in *Chapter 10*.	
M10 nuts for securing the rods of the laptop riser stand to the side frames in *Chapter 8*.	

If you are using the digital version of this book, we advise you to type the code yourself or access the code from the book's GitHub repository (a link is available in the next section). Doing so will help you avoid any potential errors related to the copying and pasting of code.

Although a programming background is not necessary, having a bit of programming experience will be very helpful.

Download the example code files

You can download the example code files for this book from GitHub at `https://github.com/PacktPublishing/Simplifying-3D-Printing-with-OpenSCAD`. If there's an update to the code, it will be updated in the GitHub repository.

We also have other code bundles from our rich catalog of books and videos available at `https://github.com/PacktPublishing/`. Check them out!

Conventions used

There are a number of text conventions used throughout this book.

`Code in text`: Indicates code words in text, database table names, folder names, filenames, file extensions, pathnames, dummy URLs, user input, and Twitter handles. Here is an example: "For our initial shape, we will subtract a circle from a square using the `difference` operation."

A block of code is set as follows:

```
difference()
{
    translate([100,0])square(200, true);
    translate([200,0])circle(80);
}
```

Any command-line input or output is written as follows:

```
module_name(parameters)
{
    body_of_module
}
```

Bold: Indicates a new term, an important word, or words that you see onscreen. For instance, words in menus or dialog boxes appear in **bold**. Here is an example: "To do so, open OpenSCAD and click on the **New** button."

> **Tips or Important Notes**
> Appear like this.

Get in touch

Feedback from our readers is always welcome.

General feedback: If you have questions about any aspect of this book, email us at customercare@packtpub.com and mention the book title in the subject of your message.

Errata: Although we have taken every care to ensure the accuracy of our content, mistakes do happen. If you have found a mistake in this book, we would be grateful if you would report this to us. Please visit www.packtpub.com/support/errata and fill in the form.

Piracy: If you come across any illegal copies of our works in any form on the internet, we would be grateful if you would provide us with the location address or website name. Please contact us at copyright@packt.com with a link to the material.

If you are interested in becoming an author: If there is a topic that you have expertise in and you are interested in either writing or contributing to a book, please visit authors.packtpub.com.

Part 1: Exploring 3D Printing

We will start our journey into 3D printing and design by focusing on 3D printers. We will learn a bit about the history of the 3D printer and then move on to how they work. We will investigate some of the software used in 3D printing before we print out our first model.

In this part, we cover the following chapters:

1
Getting Started with 3D Printing

One of the best-value 3D printers on the market today is the **Creality Ender 3 V2 3D printer**, offering a decently sized print bed with a sturdy aluminum frame. So popular is the Ender 3 V2 and other Ender 3 series printers that you can also find many upgrades and modifications to add; many of these may be 3D printed using the printer itself.

The history of 3D printers can be traced back to the 1980s. Early printers involved the use of lasers making patterns in liquids and powders. In 2005, the open source RepRap project was started and the era of 3D printers with spools of hard plastic filament was realized. Today, 3D printing is available for the general public, and is relatively affordable with machines such as the Ender 3 V2.

We are going to start our journey by having an overview of this printer before we level the bed – by far the most important step to get a good 3D print.

We will finish the chapter off with a discussion of the types of materials that we may print with the Ender 3 V2.

In this chapter, we will cover the following topics:

- Understanding the Creality Ender 3
- Leveling the print bed
- Materials available for 3D printing

Technical requirements

In this chapter, we get acquainted with 3D printers. To complete the hands-on portions, we will require the following:

- A recent 3D printer model, preferably the Creality Ender 3 V2.

- A Windows, macOS, or Linux machine.

- A microSD card and related card adapter for a computer.

- The images for this chapter may be found here: `https://github.com/PacktPublishing/Simplifying-3D-Printing-with-OpenSCAD/tree/main/Chapter1`.

Understanding the Creality Ender 3

Founded in 2014, Creality is a Chinese-based 3D printer manufacturer. Their products include the CR-10, CR-6, and Ender series **Fused Deposition Modeling** (**FDM**) printers. The Ender 3 series of 3D printers is arguably among the most iconic 3D printers not only for Creality but for the maker community at large. Some may view the Ender 3 series as entry level, but they are much more than that. The dependability, ease of use, and upgrade options available for the Ender 3 series printers make them a favorite with everyone from beginners to those with years of experience with 3D printing.

> **What Is Fused Deposit Modeling?**
>
> FDM is a technique of 3D printing where plastic filament stored on a roll is melted and deposited in place by a moving head. FDM may be referred to as **Fused Filament Fabrication** (**FFF**). FFF is the name used prior to the patent expiration of FDM in 2009.

In the following sections, we will learn about the Ender 3 series of 3D printers with a focus on the Ender 3 V2. Although the concepts covered do apply to other 3D printers, having an Ender 3 will make this section a little easier to navigate.

Ender 3 models

The first Ender 3 was released in 2018 and its design was open sourced a few months after. The following are versions of the Ender 3 printer, starting with the basic version.

Ender 3

Sporting a 220 mm by 220 mm by 250 mm build area, the **Ender 3** is the least expensive of the series and is considered the entry-level version. Aluminum extrusions provide the printer with a solid frame and both the print head and heated bed slide along their respective axes on v-slot wheels. The standard Ender 3 comes with a BuildTak-like sticker applied to the bed to provide adhesion for the first layer. We will discuss first-layer adhesion more in the upcoming *Leveling the print bed* section.

> **What Is BuildTak?**
>
> BuildTak is a proprietary product made by the company of the same name, a manufacturer of surfaces for use in 3D printing. The textured pre-cut sheets offer better adhesion than traditional methods such as painter's tape or glue sticks.

Ender 3 Pro

The **Ender 3 Pro** is an upgraded version of the Ender 3, though it has the same build area as the Ender 3 (220 mm by 220 mm by 250 mm) and is made with the same aluminum extrusions for the frame. The cooling fan for the main electronics board has been moved to vent underneath the printer to prevent bits of filament jamming the fan. The power supply has been upgraded and a removable magnetic flexible build plate has been added. This allows us to easily remove the build plate and "flex" off the printed part, as we can see in *Figure 1.1*:

Figure 1.1 – Magnetic flexible build plate

Although having a removable flexible build plate certainly has its advantages, the magnetic layer of the build plate is limited to temperatures of around 80 degrees Celsius. This somewhat limits the types of materials that can be printed with this machine. We will discuss the different types of materials in the upcoming *Materials available for 3D printing* section.

The biggest upgrade of the Ender 3 Pro is the wider aluminum extrusion for the *y* axis. This upgrade provides more stability to the *y* axis, resulting in better prints.

Ender 3 Max

The **Ender 3 Max** offers a 300 mm by 300 mm by 340 mm build area and a glass bed upgrade. The glass bed allows for printing with materials that require a high bed temperature for adhesion. The H-shaped base on the Ender 3 Max provides the extra stability required for a printer of this size.

Ender 3 V2

Coming with a new 109-mm (4.3-inch) HD color screen the **Ender 3 V2** is an upgrade to the Ender 3 and Ender 3 Pro. Keeping the same build area as the Ender 3 and Ender 3 Pro (220 mm by 220 mm by 250 mm), the Ender 3 V2 adds belt tighteners to the *x* and *y* axes. A small tool drawer has been added to the bottom of the machine for storing things like print nozzles, pliers, and scrapers.

In *Figure 1.2*, we can see the printer with its major parts identified:

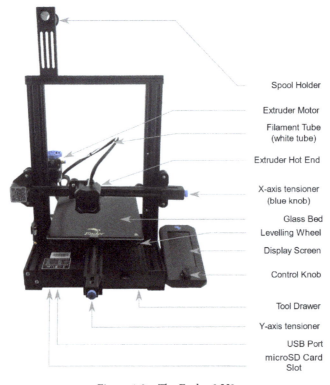

Figure 1.2 – The Ender 3 V2

We will be using the Ender 3 V2 throughout the rest of the book as our demonstration machine. The projects and descriptions using this printer that follow can not only be applied to other Ender 3 series printers but to almost all modern FDM printers on the market today.

Ender 3 S1

The **Ender 3 S1** is the latest version of the Ender 3 series. Unlike the previous versions of the Ender 3, the Ender 3 S1 comes with a direct drive extruder and built in auto bed levelling. The build area is slightly higher (220 mm by 220 mm by 270 mm) than the Ender 3 and Ender 3 V2. We will be exploring direct drive extruders in the upcoming section, *Direct drive conversion kit* where we look at upgrades for Ender 3 series 3D printers.

Understanding the parts of the Ender 3

Using *Figure 1.2* as a reference, let's take a closer look at the parts of an Ender 3 V2. The following are the major components of an Ender 3 V2 3D printer.

Spool holder

Starting from the top of the machine we have the **spool holder**. This is where we hang the spool of filament we are printing with. Spool holders can be as simple as we see in *Figure 1.2* or may be upgraded to include bearings for smoother operation. The position of the spool holder on the Ender 3 series of 3D printers has been criticized by some as the angle in which the filament enters the extruder is rather sharp. Customized upgrades such as a side spool mount (`https://www.thingiverse.com/thing:3544593`) may be added.

Extruder motor

The **extruder motor** pushes the filament through the **filament tube**, on its way to the **extruder hot end** where it is melted.

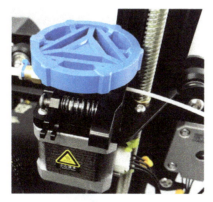

Figure 1.3 – Extruder motor

As we can see in *Figure 1.3*, the white filament on the right passes through the extruder, which is driven by a stepper motor (the black and silver part on the bottom). *Figure 1.3* is dominated by a big blue knob on the top of the extruder, used to help load the filament by hand. Turning the blue knob counterclockwise loads the filament while turning it clockwise pulls the filament out of the machine. The blue knob also acts as a visual guide that the printer is extruding during printing.

Extruder hot end

The extruder hot end is the part on the 3D printer where the filament is melted. It contains a heater block and a heat sink, which is enhanced by the use of a fan. If we were to remove the extruder hot end's case we would see that the extruder hot end looks like *Figure 1.4*:

Figure 1.4 – Extruder hot end without casing

The filament tube enters the extruder hot end through the **coupler** at the top and is pushed through to the **nozzle**. The filament is heated using a heating cartridge connected to the heater block (not shown in *Figure 1.4*). A thermistor is also connected to the **heater block** and is used to monitor the temperature (not shown in *Figure 1.4*).

In *Figure 1.5*, we see a close-up of the extruder hot end. Note the indication of the two fans, one for cooling the **heat sink** (**hot end fan**) and the other for cooling the part (**part-cooling fan**) as it is printed, as shown here:

Figure 1.5 – Extruder hot end

The part-cooling fan speed is set during the creation of the print job and can also be adjusted manually using the display screen and control knob during printing. The amount of power and thus the strength of the part-cooling fan is variable and may be changed during a print job. This is not the case for the hot end fan as it is always on full power once the Ender 3 V2 is turned on.

Filament tube

Separating the extruder motor from the extruder hot end on our Ender 3 V2 is the filament tube. Our printing material is pushed along the tube by the extruder motor to the extruder hot end, where it is melted and deposited on the bed to form our print. Designs that use a filament tube to separate the extruder motor and extruder hot end are known as **Bowden-style extrusion systems**. The Ender 3 series of 3D printers utilizes this design.

> **What is PTFE?**
>
> Filament tubes are often called PTFE tubes as they are made from **polytetrafluoroethylene (PTFE)**. PTFE was used in the 1950s to create the first non-stick cooking pans under the trade name Tefal. By being both non-stick and resistant to high temperatures, PTFE is ideal for use in 3D printer extrusion systems.

x axis and y axis tensioner

x axis and *y* axis tensioners are featured on the Ender 3 V2. They are the blue knobs at the end of their respective axes. Keeping the belts tight assists in creating better prints as the belts stretch over time.

Display screen and control knob

The biggest noticeable difference between the Ender 3 V2 and the other Ender 3 models is the screen. As we can see in *Figure 1.6*, the 109-mm (4.3-inch) color screen displays four menu options when we turn on the machine:

Figure 1.6 – Ender 3 V2 display screen

On display is the current temperature of the nozzle and bed, and the values that they are set to; as we can see, both the nozzle and bed are set to **0** degrees Celsius and are currently measuring **23** and **22** degrees for the hot end and the bed respectively. We may also see the value of the feed rate and the Z-axis offset.

The feed rate is a way of adjusting the speed of all four axes of the 3D printer (x, y, z, and extruder) together. It is adjusted during a print job to either speed up a print job or slow it down. The z-axis offset is used during printing to adjust the height of the print head relative to the bed. We may want to lower the z-axis offset if the filament is not sticking to the bed or raise it if the print head is scraping the build surface.

> **Feed Rate versus Flow Rate**
>
> Feed rate and flow rate are often confused with one another. The feed rate is controlled from the 3D printer's control panel and adjusts the speed in which the print job runs. Flow Rate controls the amount of material flowing from the nozzle and can either be set in the slicer before creating the 3D print job or adjusted during printing. We will discuss slicer programs in *Chapter 2, What Are Slicer Programs?*

Menu options are selected using the control knob. Turning the knob in one direction or another moves the selected menu option around. In *Figure 1.6*, the **Print** menu option is currently highlighted. Clicking on the knob selects the option. Please note that even though the screen may look like a touch screen, it is not.

Glass bed

Starting with the Ender 3 V2, a tempered glass bed was introduced. The tempered glass bed offers a flatter surface on which to print, compared to other bed materials. A coating added to the tempered glass bed further increases the adhesion of the filament to the bed.

Leveling wheels

Our Ender 3 V2 has four **leveling wheels** located underneath the four corners of the bed. As the name implies, these are used to level out the bed of our printer. We will use these wheels in the *Leveling the print bed* section.

USB port and microSD card slot

The USB port and microSD slot are located on the bottom left side of the machine. We use the microSD slot to load a microSD card containing our print jobs. We can also connect the printer to a computer using the USB.

> **Using a Standard SD Card**
>
> Some of us may find working with microSD cards a little troublesome due to their small size. A microSD-to-SD card extension adapter is a popular Ender 3 upgrade.

Upgrading the Ender 3

Due to the popularity of the Ender 3 series printers, many upgrades and additions exist. The following is a list of upgrades and additions that are available, but note that this is in no way a complete list.

Dual-gear extruder

A **dual-gear extruder** is a popular upgrade with the Ender 3 series printers. Adding dual gears to the extruder motor assembly adds extra grip, as we can see in *Figure 1.7*:

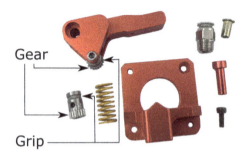

Figure 1.7 – Dual-gear extruder kit

Having the filament guided by two grips is only possible if the two sides of the rolling channel guiding the filament are synchronized or geared to each other. This extra gripping of the filament reduces skipping as the filament moves through the extruder.

As we can see in *Figure 1.7*, the spindle that is attached to the motor is geared at the bottom. When assembled, this lines up with the spindle that attaches to the arm of the extruder motor assembly. We can see that the gripping area of the motor spindle is above the geared area. When in place, this lines up with a similar grip on the arm spindle. The two spindles work in sync to pull the filament from the roll toward the extruder hot end.

Dual-gear extruder kits are relatively inexpensive and will assist in eliminating filament slippage. Please note that we must change the extruder motor steps per mm setting after installing a dual-gear extruder.

Nozzles

Typically, 3D printers come with a standard brass nozzle with a 0.4-mm hole; however, nozzles with different diameters may be purchased rather inexpensively. Nozzle hole diameters come in a variety of sizes, including 0.2 mm, 0.3 mm, 0.4 mm, 0.5 mm, 0.6 mm, 0.8 mm, and 1.0 mm.

Smaller hole diameters increase the printing time but produce more detailed prints. Larger nozzle holes reduce the printing time but at the cost of quality.

3D printer nozzles are typically made of brass. Brass offers excellent heat transfer for the price. Brass nozzles do tend to wear quickly however and are not well suited to materials that are a little rougher in texture such as wood and carbon fiber. For such materials, stainless steel and hardened steel nozzles are desired. In *Figure 1.8*, we can see from left to right a 0.4-mm brass nozzle, a 0.4-mm stainless steel nozzle, and a 0.6-mm hardened steel nozzle:

Figure 1.8 – Various 3D printer nozzles

As we see in in *Figure 1.8* nozzles may have different thread sizes. For our Ender 3 V2, we need to use nozzles with M6 threading.

Direct drive conversion kit

Like many 3D printers on the market, the Ender 3 V2 comes equipped with a Bowden tube style extrusion system. To understand what exactly this is, let's look at the left-hand diagram in *Figure 1.9*:

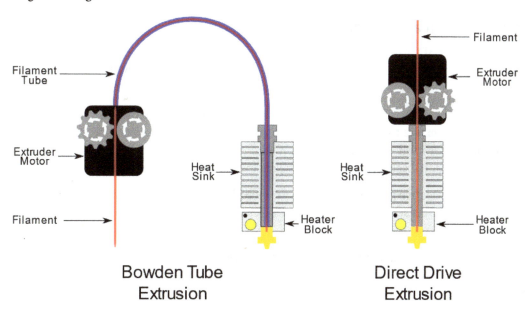

Figure 1.9 – Bowden tube extrusion versus direct drive extrusion

With **Bowden tube extrusion**, the filament is pushed through the PTFE tube (Filament Tube) into the heater block through the heat sink, where it is melted and deposited onto our printer bed. By contrast, as we can see in the right-hand diagram, a **direct drive** extrusion system pulls the filament toward the heater block through the heat sink to the nozzle.

With Bowden tube extrusion, the extruder motor is mounted separately from the other parts. With direct drive extrusion, the extruder motor is mounted with the heat sink and heater block.

Direct drive extrusion kits for the Ender 3 series printers are not particularly expensive and may be installed with relative ease. We will be using the stock Bowden tube extrusion setup for the projects in the book.

Is Direct Drive Better Than Bowden?

The debate as to which system (direct drive vs. Bowden) is better can be a heated one in the maker community. In a Bowden tube extrusion system, the print head (including the heat sink, heater block, and nozzle) moves more quickly than in a direct drive extrusion system due to its lighter weight (as the extruder motor is separate). However, direct drive extrusion systems tend to work better with flexible materials than Bowden tube extrusion systems as it is easier to pull a flexible filament into the heater block than it is to push it through a tube.

OctoPrint with a Raspberry Pi

Another popular 3D printer upgrade is **OctoPrint**. Using a Raspberry Pi connected to our 3D printer, we can use OctoPrint to run and monitor print jobs remotely. This includes hooking up a USB camera for video monitoring.

We can upgrade our OctoPrint setup with items such as OctoDash to provide a touchscreen interface to OctoPrint. With OctoDash, the 3D printer can be controlled right at the printer itself. Other additions to OctoPrint include the Enclosure plugin, which uses additional sensors to monitor the enclosure the printer may be in.

The Spaghetti Detective plugin and service for OctoPrint provides AI monitoring for our prints. An alert is sent when the Spaghetti Detective service determines that a print has failed.

Alternatives to OctoPrint include AstroPrint and Repetier-Server.

Tent enclosure

Arguably one of the best additions we can make to our 3D printer setup is an enclosure such as a tent. Tent enclosures are constructed like a tent used for camping and usually have more than enough room to fit our Ender 3 V2.

Enclosures allow a consistent temperature, resulting in better print quality. Enclosures are perfect for 3D printers that are used in garages. Adding a wireless dehumidifier inside the enclosure will help keep the humidity down and will assist in printing with filaments that are susceptible to retaining moisture. Tent enclosures also offer a layer of protection as they are generally fireproof.

In *Figure 1.5*, we can see an Ender 3 V2 inside a tent enclosure.

All-metal hot end

If we were to take the hot end of our stock Ender 3 V2 apart, we would see that the PTFE tube extends all the way to the nozzle. This is illustrated in the left-hand diagram in *Figure 1.10*:

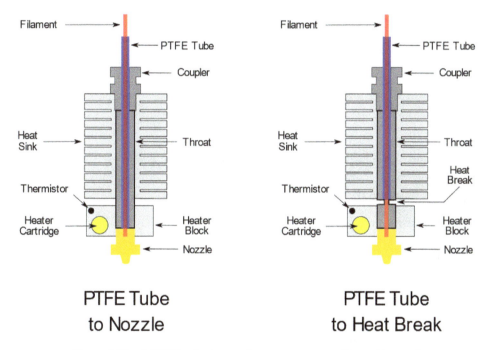

Figure 1.10 – A PTFE-tube-to-nozzle setup versus an all-metal hot end

This type of setup works well for materials with lower melting points such as PLA as temperatures above 230 degrees Celsius or so will start to melt the PTFE tube, causing blockages.

For higher-temperature materials such as ABS, an all-metal hot end is desired. As illustrated on the right in *Figure 1.10*, with an all-metal hot end the PTFE tube ends at the heat break, where the filament continues through to the heater block without the PTFE tube.

Bi-metal heat break

An upgrade for the all-metal hot end is a bi-metal heat break. The bi-metal heat break is made up of two separate metals, a stainless-steel inner tube, and a brass outer tube. We can see a picture of a bi-metal heat break in *Figure 1.11*:

Figure 1.11 – Bi-metal heat break

Due to the poor temperature transfer from the thin inner stainless-steel tube to the outer brass, the bi-metal heat break keeps higher temperatures from creeping up the extruder hot end to the heat sink. This allows for faster extrusions and the ability to print with higher-temperature materials.

PEI build plate

PEI (or polyetherimide) build plates offer an excellent alternative to existing build plates. They are relatively maintenance-free, only requiring cleaning with isopropyl alcohol. Their flexibility makes it easy to "flex" a part off the build plate.

PEI build plates have great adhesion properties for 3D printer filament. Some PEI build plates for the Ender 3 series are two-sided, with a smooth side and textured side (for creating prints with a textured bottom layer).

Auto bed leveling sensor

Bed leveling for a stock Ender 3 series printer involves adjusting the leveling wheels under the print bed. Another option is to automate the process with an auto bed leveling sensor such as a BL Touch. In this book, we will be leveling our bed manually and will not be installing an auto leveling system.

Capricorn tubing

One upgrade for our PTFE filament tube is to use Capricorn PTFE tubing. Capricorn started out in 2016 with the goal of producing the best Bowden-style tubing. Capricorn tubing generally comes in blue and has a higher temperature rating than typical PTFE tubes.

DIY upgrades

Due to the popularity of the Ender 3 series printers, there are many DIY upgrades available to 3D print. In fact, the small tool drawer that exists on the Ender 3 V2 was adapted from DIY Ender 3 drawers found online. Websites such as `www.thingiverse.com` and `www.myminifactory.com` offer many 3D files of Ender 3 upgrades that we can download and print ourselves.

Now that we are more familiar with the Ender 3, let's perform the most necessary task for ensuring that our print jobs are successful – leveling the bed.

Leveling the print bed

Arguably the most important thing we can do to ensure high-quality 3D prints is to properly level the print bed. In this section, we will manually level our print bed by moving the extruder hot end to each corner and adjusting the bed using the leveling wheels.

Before we level the bed on our 3D printer, we should take note of the importance of having a perfectly flat build surface sitting on top of the bed. Having an uneven build surface makes the adhesion of the first layer difficult. Choosing the right build plate material will make the task of leveling out the bed much easier.

Glass is an extremely popular build surface due to its flatness. Borosilicate glass is often used for build surfaces due to its thermal properties as it can withstand great temperature variations without cracking.

Leveling the corners of the bed

Leveling the corners on the bed is the easiest way to level the print bed with relation to the nozzle. To ensure that this works, the surface of the build plate material (the glass plate for the Ender 3 V2) must be perfectly flat.

To begin the process, we will use the control panel again. We will first set the print head (extruder hot end) to the home position and then move it around the build plate. The following instructions are for the Ender 3 V2, while other Ender 3 printers or 3D printers with Marlin firmware will work similarly:

1. Prepare a small rectangular piece of paper about 10 cm by 5 cm.
2. Scroll to the **Prepare** menu and click on it. For other Ender 3 models, click on the control knob, navigate to the **Prepare** menu, and click on it.
3. Navigate to the **Auto Home** menu option and click on it. Observe that the print head moves to the home position.

4. Scroll to the **Move** menu option and click the control knob to select it.

5. Scroll down to the **Z** menu option, click the control knob to select it, and dial in the value 20 by turning the dial clockwise. Click to set it. Observe that the print head moves up 20 mm.

6. Scroll to the **X** menu option and click to select.

7. Set the value to 20mm and click to set it. Observe that the print head moves 20 mm in the X direction.

8. Scroll to the **Y** menu option and click to select.

9. Set the value to 20mm and click to set it. Observe that the print head moves 20 mm in the Y direction.

10. Slide the piece of paper under the print head:

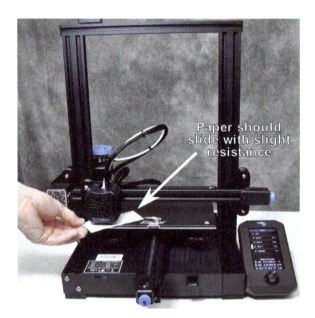

Figure 1.12 – Bed leveling

11. Scroll down to the **Z** menu option, click the control knob to select it, and dial in the value 0 by turning the dial clockwise. Click on the control knob to set this. Observe that the print head moves down and touches the piece of paper.

12. Using the leveling wheel closest to the point on the bed, turn the wheel to either lower or raise the bed so that the paper can move freely under the print head with a slight tug. The paper should not rip, nor move freely without a little bit of resistance (*Figure 1.12*). Use the graphic in *Figure 1.13* to determine how to move the bed either up or down:

Bed Position

Figure 1.13 – Adjusting the bed position

13. Repeat *steps 6 to 12* with an *x* value of 180 and *y* value of 20.

14. Repeat *steps 6 to 12* with an *x* value of 180 and *y* value of 180.

15. Repeat *steps 6 to 12* with an *x* value of 20 and *y* value of 180.

16. Set the *z* axis to 20mm.

17. Select **Auto Home** to home the printer.

We have just leveled the bed by manually leveling the corners.

Mesh bed leveling

Our print bed should be leveled and ready to print after leveling the corners. However, in cases where it dips or rises between the corners, we have a few options we could apply to address this, as follows:

- Replace the build surface with a new one.

- Print using rafts (we will investigate this in *Chapter 3, Printing Our First Object*).

- Install mesh bed leveling on our Ender 3 V2.

Buying a new build surface like a new glass bed or PEI plate is an easy option to take as our build plates do get worn with use. There are many build surfaces for the Ender 3 to choose from.

However, if that option is not available, we can do what used to be common prior to glass beds, which is to print our parts on rafts. Basically, rafts are flat surfaces that are printed onto our bed before printing our part (a raft for the part, so to speak). Rafts fell out of favor when glass beds became popular, as rafts can sometimes be difficult to remove from the part and they waste precious material that will only be thrown out.

The third option we can explore (for the Ender 3 V2 only) is mesh bed leveling:

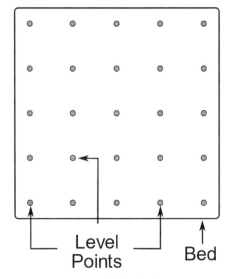

Figure 1.14 – Mesh bed leveling

As we can see in *Figure 1.14*, mesh bed leveling involves taking measurements at many points on the bed. These values are then used to calculate where to set the z axis on the print head as it moves around the bed.

To get mesh bed leveling on our Ender 3 V2 we must update the firmware. The firmware is the program that runs on the controller board of our 3D printer. The firmware may be updated in a couple of ways:

- By loading pre-compiled firmware onto a microSD card and installing it on our Ender 3 V2

- By loading and compiling the firmware source code using a program such as Arduino IDE

> **A Few Good Reasons to Update the Firmware**
>
> Upgrading the firmware on our Ender 3 V2 will give us extra features in addition to mesh bed leveling. Scrolling text for long filenames is added, as well as the ability to load files from subfolders and not just the root. A new main menu option called **Level** pushes **Info** to the last option under the **Control** menu. Also, a white border has been added when selecting menu options, making the main menu easier to see.

We will use the first option to install mesh bed leveling on our Ender 3 V2.

Updating the firmware

We can find a list of Ender 3 V2 firmware with mesh bed leveling at the following GitHub repository: `https://github.com/Jyers/Marlin/releases`.

To know which version of the firmware to download, we need to find out which board is installed in our Ender 3 V2 (please note that this is an upgrade for the Ender 3 V2 and isn't available with previous Ender 3 printers). To do this, complete the following steps:

1. Remove the cover of the main controller board by removing one screw from the top and three screws from the bottom.

2. Place the Ender 3 V2 on its side and observe the board number. See *Figure 1.15* to find this:

Figure 1.15 – Finding the board number on our Ender 3 V2

3. For our Ender 3 V2 we can see that the board number is V4.2.2. Download the appropriate `.bin` file from the website listed at the beginning of this section. For our printer, we will download the `E3V2-ManualMesh-5x5-v4.2.2.bin` file.

4. Be sure to download the correct `E3V2-ManualMesh-5X5` file for the board version. We will be calibrating a 5x5 mesh. Load the `.bin` file onto a formatted microSD card.

5. Ensure that the Ender 3 V2 is turned off. Load the microSD card into the printer. Turn on the Ender 3 V2 and observe that the firmware has been updated. The **Info** menu option should be replaced by the **Level** menu option. There should be x, y, and z values for the print head displayed at the bottom of the display screen.

We are now ready to level the bed using mesh bed leveling.

Running mesh bed leveling

With the firmware installed, mesh bed leveling involves taking z-axis measurements at 25 points on the bed. At the end of the process, the mesh is saved and used when we 3D print.

To level our bed with mesh bed leveling, do the following:

1. From the main menu navigate to the **Level** menu option and click the control knob.

2. Observe that the print head moves to the home position before moving to the first mesh point.

3. Slide a 10cm by 5cm piece of paper under the print head. The paper should slide under the print head with a slight tug.

4. If the paper does not slide under the print head with a little resistance, then we must raise or lower the print head. Using the **Microstep Up** and **Microstep Down** options, adjust the print head accordingly by pressing and holding the control knob. Please note that up and down are opposite of what they were when we were leveling the print bed with the leveling wheels.

5. Scroll up and click on the **Next** menu option when satisfied with the level. Observe that the print head moves to the next position.

6. On the last measurement point there will be a menu option called **Save Mesh**. Scroll up to this menu option and click the control knob to save the mesh. Observe a double beeping sound indicating that the mesh has been saved.

7. Ensure that the **Leveling Active** check box from the **Level** menu is checked.

We have now successfully leveled our print bed using mesh bed leveling.

Automatic mesh bed leveling

Some of us may have noticed a file called `E3V2-BLTouch-5x5-v4.2.7.bin` when we were downloading the firmware. BL Touch is an after-market sensor that we can add to our Ender 3 V2 series printer. We would use firmware such as this if we had an automated leveling sensor such as BL Touch installed on our printer.

> **How Often Do We Need To Level The Bed?**
>
> The bed of our 3D printer should not need leveling very often if we take care not to apply too much pressure to the bed when removing prints. The most common mistake many make is not letting the bed cool down to room temperature before removing the printed part. In most cases, the printed part will just slide off the glass bed once it has returned to room temperature.

Now that we have leveled our print bed, let's take a look at some of the material available for use in 3D printing.

Materials available for 3D printing

It used to be that there were essentially only two materials available for 3D printing, **Poly-Lactic Acid** (**PLA**) and **Acrylonitrile Butadiene Styrene** (**ABS**). This has changed considerably over the last few years. The result has not only given us new materials with which to 3D print; it has changed what we can make with our 3D printers.

Materials used for FDM 3D printing come in the form of a rolled plastic filament of either 1.75 mm or 2.85 mm in diameter. A spool made of plastic, cardboard, or metal is used to hold the filament. Spoolless filament for installing on a reusable spool is also available.

Let's look at the materials we can use for 3D printing, starting with PLA and ABS.

Poly-Lactic Acid (PLA)

PLA is the most used material for 3D printing. It is made from sugar cane or corn starch and is biodegradable, making it an eco-friendly option. Compared to many other filament materials, PLA is extremely easy to work with.

Although not requiring a heated bed (many early 3D printers did not have heated beds), PLA does benefit greatly from heat applied to the bed due to its low melting temperature, which increases its stickiness.

Figure 1.16 shows a part for a small desktop monitor table printed in red PLA:

Figure 1.16 – A part printed on the Ender 3 V2

Early PLA was quite brittle, making it unsuitable for many applications, but in recent years PLA has got a lot better, especially in terms of its strength.

When printed at around 200 degrees Celsius at the hot end and 60 degrees Celsius on the bed, PLA provides an excellent finish (temperature may vary with manufacturer).

PLA can be glued with epoxy, providing opportunities to break up larger objects into smaller parts.

Acrylonitrile Butadiene Styrene (ABS)

ABS is another common material for 3D printing and is a popular plastic for making toys. LEGO blocks, for example, are made from ABS.

In *Figure 1.17* we see the cat figurine printed in ABS using the G-code file that comes with the Ender 3 V2:

Figure 1.17 – Cat figurine printed with ABS

ABS prints with a certain smell that many find unpleasant and thus printing in a separate room is encouraged.

ABS produces prints that are more durable than PLA and with a higher melting temperature. This makes it more ideal in situations where a part may be subjected to higher temperatures. Having ABS stick to the print bed can be challenging. A heated bed is necessary to produce prints that do not warp upward at the edges and stay flat on the bed throughout the print job. ABS should be printed with a nozzle temperature of around 240 degrees Celsius and a bed temperature of 90 degrees Celsius.

An enclosure is encouraged with ABS printing to avoid cooling cracks on the part during printing.

ABS prints can be smoothed with acetone to hide the layer lines. As we can see in *Figure 1.17*, layer lines around the top of the print are noticeable. Subjecting our print to an acetone vapor bath will melt the lines together, resulting in a smooth professional-looking 3D print.

An alternative to ABS is ASA, which has similar properties but with UV protection. This makes ASA well suited for outdoor applications.

Glycolyzed Polyester (PETG)

PETG is a modified version of **PET** (or **polyethylene terephthalate**), a plastic used extensively in the production of water bottles and food containers since the 1990s. Adding glycol to PET to reduce its brittleness turns it into PETG.

PETG filament prints almost as easily as PLA and provides the strength of ABS with a lower melting temperature. It is known for its impact resistance, light transmission (when transparent filaments are used), and its food-contact safety approval. In *Figure 1.18*, we see two parts that make up a clamp for a CNC router:

Figure 1.18 – CNC router part made with two different materials

The part on the right was printed with black PETG. PETG parts tend to be shinier and less brittle than PLA. PETG works well for functional parts. It is exceedingly difficult to glue PETG parts together so other construction techniques, such as incorporating nuts and bolts, must be used.

> **PETG and Glass Beds**
> PETG should not be printed directly to a glass bed as it sticks a little too well. Removing it from a glass bed can result in chips to the glass and possibly an expensive repair.

High-Impact Polystyrene (HIPS)

HIPS has similar properties to ABS but is lighter in weight. HIPS dissolves easily in d-limonene, making it an ideal dissolvable support material for ABS and other materials. A 3D printer with a dual extrusion system (see *Figure 1.19*) is required when using a different material such as HIPS as a support material, as shown here:

Figure 1.19 – Dual extruder 3D printer

> **Dual Extruder 3D Printers**
>
> Dual extruder 3D printers (not to be confused with dual-gear extruders) use two extruders that move together along the y and z axes but opposite to each other on the x axis. For dissolvable support prints, one extruder extrudes the support material, such as HIPS or PVA, and the other extruder delivers the material with which we want to print. Dual extruder 3D printers are also referred to as IDEX 3D printers.

Printing with HIPS requires a heated bed and should be done in a separate room due to the fumes. When not used as a support material, parts made with HIPS tend to be lightweight and rigid and may be easily sanded and painted.

Polyvinyl Alcohol (PVA)

PVA is to PLA what HIPS is to ABS, a dissolvable support material. In the case of PVA, however, it is dissolvable in warm water. PVA is very hygroscopic (meaning it absorbs moisture) and must be as dry as possible when printed with.

PVA requires a heated bed set to a temperature of around 60 degrees Celsius. PVA is printed with a nozzle temperature between 190 – 210 degrees Celsius. When used with dual extruder 3D printers, the extruder loaded with PVA should have its heater block turned off when not in use so as to avoid jamming.

Carbon fiber

Carbon fiber is used to lace other filament types, such as PLA, ABS, nylon, and PETG, to make them stronger. Prints made with a carbon fiber-laced filament can be made lighter due to this additional strength. It should be noted, however, that carbon fiber is abrasive to the nozzle on our extruder and thus a hardened steel nozzle is recommended when printing with carbon fiber-laced filaments. In *Figure 1.20*, we can see an arm for a quadcopter 3D printed with a carbon fiber-laced PETG filament:

Figure 1.20 – Quadcopter arm

This part weighs just 8.5 grams and cannot be bent by hand.

Nylon

Nylon is one of the toughest plastics available. It is used in many products, such as zip ties. 3D printing with nylon can be challenging as it is a hygroscopic material and must be printed dry. Nylon prints are tough but slightly flexible. *Figure 1.21* shows a 3D-printed nylon replacement buckle for a hockey helmet:

Figure 1.21 – 3D-printed buckle

The buckle flexes enough to be pushed over the metal button on the helmet.

Dry Boxes

Many materials used in 3D printing are very hygroscopic, meaning they absorb moisture from the air. This causes issues when printing as their diameters swell and jam up the extruder. A solution to this issue is to print from a dry box. A dry box is an airtight storage container with a small hole where filament is passed through and fed to the extruder on a 3D printer. Dry boxes may be purchased or easily made from existing airtight containers. There are also many DIY designs for dry boxes at places such as Thingiverse.com.

Flexible materials

Flexible filament can be used to print things such as phone cases and gaskets. In *Figure 1.18*, the part on the left was printed with a flexible material called **NinjaFlex**.

The term **Thermoplastic Elastomer** (**TPE**) is used to describe the blend of elastic and thermoplastic (soft rubber and hard plastic) that makes up the flexible material we 3D print with. **Thermoplastic Polyurethane** (**TPU**) is a common type of TPE that is more on the rigid side.

To better understand flexible filament and its uses, it is good to understand the **Shore hardness scale**. Comprising measurement devices (called durometers) calibrated at different strengths, there are three main scales: Shore OO, Shore A, and Shore D. In *Figure 1.22*, we can see the hardness of common non-metallic items:

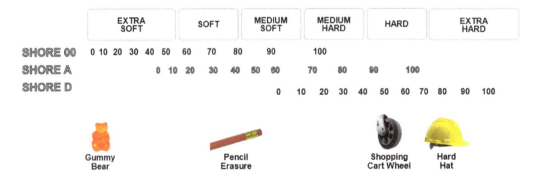

Figure 1.22 – Shore hardness scale

From *Figure 1.22*, we can see that a shopping cart wheel has a Shore hardness of 95A or 50D. The NinjaFlex used in *Figure 1.18* has a hardness of 85A, meaning it is harder than a pencil eraser but softer than a shopping cart wheel.

Bowden tube extrusion systems tend to struggle at printing with flexible materials. Generally, a direct drive extrusion system is used to 3D print with flexible materials.

Other materials

Other materials available for 3D printing include **wood-laced filaments**, **polycarbonate filaments**, **metal filaments**, **PolyEtherEtherKetone (PEEK) filaments**, and so on. Many of the high-performance filaments require industrial 3D printers with heated chambers and could not be printed with an Ender 3 series printer.

As we can see, there are many materials available for use with 3D printing, and they are getting better and stronger all the time. In this book, we will work mainly with PLA and ABS.

Summary

In this chapter, we discussed the Ender 3 series range of 3D printers, with a particular focus on the Ender 3 V2. We looked at the major components of the Ender 3 V2 and described some of the upgrades and additions we can add to it.

In the hands-on section of this chapter, we leveled the bed on our Ender 3 V2, by far the most important step toward quality 3D prints from a 3D printer. We explored upgrading the firmware on our Ender 3 V2 to get access to the mesh bed leveling functionality, as well as some other upgrades included in the new firmware.

We closed off the chapter by looking at the various materials we can 3D print with, including a look at the Shore hardness scale in order to understand flexible materials in a bit more depth.

In the next chapter, we will look at the software used to create print jobs for our 3D printer on our way to bringing our 3D design ideas to life.

2

What Are Slicer Programs?

To 3D-print an object, we must first create instructions on how to do so in a language that the 3D printer understands. This language is called **G-code**, and to describe it in its simplest form, it is code that tells the printer head where to move and when. To create G-code, we utilize software called slicers.

In this chapter, we will create simple G-code programs before we investigate the various slicer programs in use today for 3D printing.

In this chapter, we will cover the following topics:

- Controlling a 3D printer using G-code
- Common FDM slicer programs
- Slicer programs for liquid resin 3D printers

Technical requirements

In this chapter, we will get acquainted with 3D printers. To complete the hands-on portion, we will require the following:

- A late-model 3D printer, preferably the Creality Ender 3 V2.

- A Windows, macOS, or Linux computer with a USB cable.

- The code and images for this chapter can be found here: `https://github.com/PacktPublishing/Simplifying-3D-Printing-with-OpenSCAD/tree/main/Chapter2`.

Controlling a 3D printer using G-code

Computer Numeric Control (**CNC**) is a method of controlling a machine from a computer or controller. Early CNC machines from the 1940s used punch tape and were used to crudely control machines of the time. Combining advanced computer systems with machines in the 1960s gave way to the CNC machine we know today.

3D printers are, in essence, a form of CNC machine. While CNC machines are subtractive as they chip away material to make parts, 3D printers are additive as they deposit material. G-code is the language that CNC machines and 3D printers use to communicate with their respective controllers.

In this section, we will explore G-code and use it to control our printer.

What is G-code?

So, what exactly is G-code? As mentioned, G-code is the language that 3D printers and CNC machines use for instructions. To get a more detailed understanding, let's look at how a computer communicates with a 3D printer.

Looking at *Figure 2.1*, we can see that the computer sends G-code commands to the 3D printer and receives sensory data back:

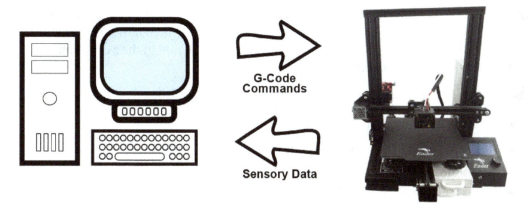

Figure 2.1 – Communicating with a 3D printer

Such G-code commands may be used to home the printer or set the temperature of the hot end or bed. Sensory data coming back can be the hot end temperature, the bed temperature, or an indication that one of the limit switches for an axis (x, y, and z) has been engaged.

To gain a deeper understanding of G-code, let's break down G-code commands.

Understanding G-code

G-code for a 3D printer consists of commands that start with the letter G or the letter M and control the movement and functionality of the 3D printer.

In *Figure 2.2*, we can see a list of G-code statements on the left. If we were to execute each statement sequentially, the print head would be positioned in the spot we see in *Figure 2.2*:

Figure 2.2 – G-code commands used with a 3D printer

The first statement, G28, homes the print head on all axes. The G0 Z10 command moves the print head to 10 mm above the bed. G0 X120 Y120 Z20 then moves the print head 120 mm in the *x* direction, 120 mm in the *y* direction, and 20 mm in the *z* direction (the position as shown in *Figure 2.2*).

The following are some common G-code commands:

- G0 – fast linear motion
- G01 – controlled linear motion set by an additional F parameter (for example, G1 Z15.0 F9000)
- G28 – auto home
- M104 – set the hot end temperature
- M140 – set the bed temperature
- M117 – set the LED message
- M106 – set the cooling fan speed

By stringing together G-code commands in a file, we can have our 3D printer perform tasks such as executing print jobs and bed leveling.

Using Pronterface to control our 3D printer

Pronterface is a 3D printer control and printing application written in Python. Although a little dated, it offers a simple GUI in which to control our 3D printer. To install Pronterface onto our computer, follow these steps:

1. Navigate to http://www.pronterface.com.
2. Click on the **DOWNLOAD** link.
3. Find the appropriate installation file, and then download and install it (please note that the program may be referred to as Printrun).

Once installed, open up Pronterface. We can use Pronterface for either the Ender-3 V2 or the other 3D printers. To connect the computer to our printer, do the following:

1. Connect the Ender-3 V2 or another 3D printer to the computer using a USB cable.
2. From the **Port** selection at the top left, select the proper port for the printer.
3. Set the baud rate to **115200** (this value may be dependent on the 3D printer).
4. Click on the connect button and observe that the printer connects to the computer.

Now that we have our printer connected, let's take a short look at the GUI. Observe the control wheel at the top left-hand side of the program, as shown in *Figure 2.3*:

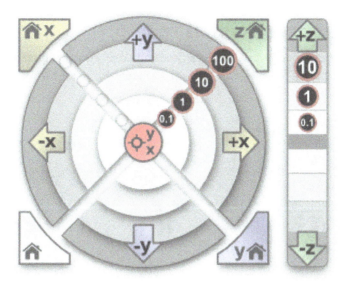

Figure 2.3 – The Pronterface control wheel

To verify that we can control our printer with Pronterface do the following:

1. Click on **10** under the **+Z** button and observe that the print head moves up 10 mm in the **z** direction.

2. Click on the **x** homing button at the top left and observe that the print head moves left and stops at the **x** home position.

Now that we have verified that we can control our printer using Pronterface, let's start writing some G-code. At the bottom right corner of Pronterface, observe a textbox with a button named **Send** beside it.

Type the following into the textbox and click on **Send**:

```
G28
```

Observe that the print head is moved to the home position for all axes (*x*, *y*, and *z*).

> **The G28 Command**
>
> The G28 command on its own will home all axes as we have observed. To home an axis by itself, put the axis name after G28 – for example, to home the *y* axis, type in G28 Y.

You may recall from *Chapter 1*, *Getting Started with 3D Printing*, how tedious manually leveling the corners of our print bed was, so let's write some G-code to assist us.

Leveling the corners with G-code

Putting our G-code knowledge to work, let's create a bed leveling program. In a text editor (such as Notepad in Windows), type in the following and save it with the filename `level-bed.gcode`:

```
G28
G0 Z20
G0 X20 Y20
M0 Position paper
G28 Z
M0 Adjust level
G0 Z20
G0 X180 Y20
M0 Position paper
G28 Z
M0 Adjust level
G0 Z20
G0 X180 Y180
M0 Position paper
G28 Z
M0 Adjust level
G0 Z20
G0 X20 Y180
M0 Position paper
G28 Z
M0 Adjust level
G0 Z20
G28
```

Be sure to use the `.gcode` extension in the filename. Before we run our program, let's look at some of the commands we have just written.

We start the program with `G28`, which we know homes the print head. From there, we move the print head up 20 mm (`G0 Z20`), then 20 mm in the *x* direction (`G0 X20`), and 20 mm in the *y* direction (`G0 Y20`).

The `M0 Position Paper` command creates a pause and displays the **Position Paper** message on the LCD screen of the printer. The printer will stay paused until we click on the control knob.

It is time to run our program in Pronterface. To complete this, follow these steps:

1. From Pronterface, click on the **Load file** button located at the top center of the screen.
2. Locate `level-bed.gcode` and load it by clicking **Open**.
3. Click on the **Print** button at the top center of the screen.
4. Observe that the print head is homed before moving to the first leveling position. We should see the **Position paper** message on the LCD screen. Place a piece of paper the same size as we used in *Chapter 1*, *Getting Started with 3D Printing*, under the print head.
5. Click on the control knob. Observe that the print head lowers to the print bed. Observe the **Adjust level** message on the LCD screen.
6. Referring to *Figure 1.13* from *Chapter 1*, *Getting Started with 3D Printing*, adjust the nearest leveling wheel so that there is a slight tug on the paper. The paper should not rip.
7. Click on the control knob to move to the next leveling point.
8. Repeat until all four corners have been leveled.

> **Running Our Code from the 3D Printer**
>
> We actually do not need a computer connected to our 3D printer to run our `level-bed.gcode` program. We could just simply load it onto a microSD card and run it from the 3D printer. Running print jobs from the 3D printer itself is the most common way to run a print job, as it does not tie up a computer for hours.

Congratulations are in store, as we have just written and run our own bed leveling program using a little bit of G-code! It is now time to look at slicer programs that create G-code for our 3D printers.

Common FDM slicer programs

As we discussed in the previous section, G-code is the language used to control 3D printers. G-code controls both the movement of the print head and the extrusion of plastic from the nozzle, which allows us to create physical objects with our 3D printer.

But how do we create the G-code needed to print an object? Writing the G-code ourselves is obviously an exceedingly difficult thing to do. This is where slicers come in.

Before we explore a few slicer programs available for our FDM 3D printer, let's look at what a slicer does.

Slicing an object into G-code

A slicer is software that takes 3D object files and converts them into G-code that our 3D printer understands.

In *Figure 2.4*, we can see the process documented graphically:

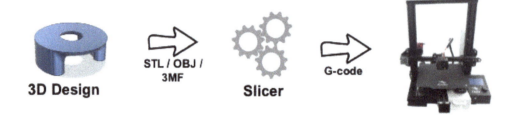

Figure 2.4 – Converting a 3D object file into G-code

A 3D design – in this case, a riser for a computer monitor – designed in **Computer-Aided Design (CAD)** software is converted to a 3D object file. 3D object files are generally stored as STL, OBJ, or 3MF files, and are created using CAD software.

The slicer analyzes the 3D object file and "slices" it into layers, divided along the z axis. Each layer is a series of G-code commands for the x axis, y axis, and the extruder. The number of layers is dependent on the layer height set in the slicer software.

In *Figure 2.5*, we can see the first and last slices of our computer monitor riser:

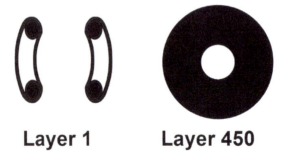

Layer 1   Layer 450

Figure 2.5 – The first and last layer slices of a 3D-printed job

As our object is 90 mm tall, when it is sliced with a 0.2 mm layer height, the result is G-code with 450 different z axis values or layers. Another way to look at it would be to visualize the print head moving up 0.2 mm on every layer 450 times during printing.

As we can see, software to perform the slicing process is valuable. Let's look at the various software applications available to do this.

Slicing software applications

Many people confuse the software applications used to prepare and control a 3D printer with the slicing software itself. For example, Pronterface has functionality to prepare a 3D object file from the slicing stage to running the print job on the 3D printer. We can also control our 3D printer with Pronterface. However, for slicing functionality, Pronterface uses the program Slic3er, which itself is offered as a standalone program.

The following are three of the available software programs we can use to create the G-code that our FDM 3D printer requires.

Slic3r

As mentioned, **Slic3r** is used as the slicing engine in other programs such as Pronterface and Repetier-Host. As a standalone program, Slic3r is incredibly powerful, offering control over a vast number of parameters.

Slic3r is available for Windows, macOS, and Linux and can be found at `www.slic3r.org`. In *Figure 2.6*, we can see what Slic3r looks like once we open it up and load a file to be processed, with a few areas highlighted:

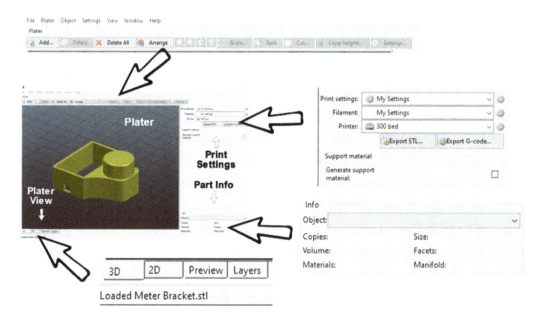

Figure 2.6 – The Slic3r main page

As we can see in *Figure 2.6*, the plater takes up the most space on the screen. A mount for a hygrometer sits on the build plate.

Slic3r is considered fast at slicing. The following are some interesting features of Slic3r:

- The default settings in the **Print | Settings** section make it easy to create G-code for our 3D printer. Clicking on the **Export G-code** button in this section creates a G-code file to load into our 3D printer. Slic3r is not used for 3D printer control.

- Slic3r can change layer heights for part of a print. For example, our print job may start with a 0.2 mm layer height and change to a 0.1 mm layer height later in the print job. This is useful in situations where the top of a print may be rounded and smaller visible layer lines are desirable.

- Another feature worth noting is the ability of Slic3r to create a series of SVG (vector) images using the **Slice to SVG...** option under the **File** menu. At one time, this functionality was one of the few ways to slice an object for SLA (Stereolithography or liquid resin printing). However, with the plethora of slicer programs for liquid resin printers today, the **Slice to SVG...** feature is not as useful as it once was.

ideaMaker

ideaMaker by Raise3D is an advanced slicing and 3D printer control program. Built specifically for Raise3D's brand of industrial 3D printers, ideaMaker can be used on 3D printers from other manufacturers, including the Ender-3 V2.

ideaMaker is feature-rich, giving us the ability to cut models into pieces, easily add supports (automatic and manual), and even repair models with non-manifold edges. ideaMaker also can upload a print job directly to OctoPrint, creating an efficient slicing to printing workflow.

To get a hands-on feel for ideaMaker, let's try out a few of these more distinctive features. In the following examples, we will demonstrate usage of the Free Cut, Repair, and Texture tools.

Preparing our model

Before we start, we need to install ideaMaker and load a model. To do this, follow these steps:

1. To download and install ideaMaker, go to `https://www.raise3d.com/ideamaker/`.

2. We need an object to load into ideaMaker. A quite common 3D printer test print is 3DBenchy. To download a 3DBenchy model, go to `http://www.3dbenchy.com/download/`.

3. To load our 3DBenchy model, use the **Add** button at the top left of the ideaMaker screen. Our 3DBenchy will be stored as a `.stl` file. For models downloaded from Thingiverse, the `3DBenchy.stl` file will be found in the `files` folder.

4. Observe that our 3DBenchy model is loaded onto the center of the build area:

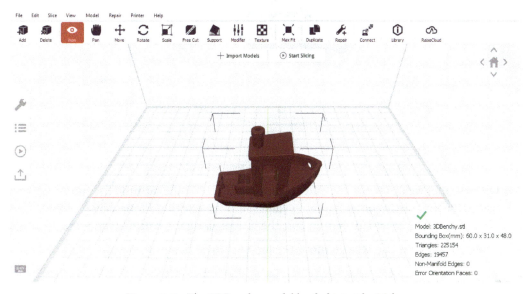

Figure 2.7 – The 3DBenchy model loaded into ideaMaker

Now that we have our 3DBenchy model loaded, it is time to cut it in half and repair the two halves.

Cutting our object

To cut our 3DBenchy into two parts, follow these steps:

1. Select the model using the left mouse button.

2. Select the **Free Cut** tool from the top toolbar:

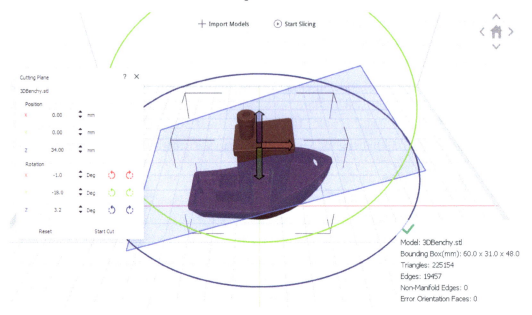

Figure 2.8 – Using the Free Cut tool in ideaMaker

3. Observe that we can move the cutting plane (blue rectangle) around the model using the orbit circles or by typing in values into the **Cutting Plane** dialog box (*Figure 2.8*). The **Start Cut** button cuts the model along the cutting plane.

4. After the model is cut, we can move the top part to be flat on the bed:

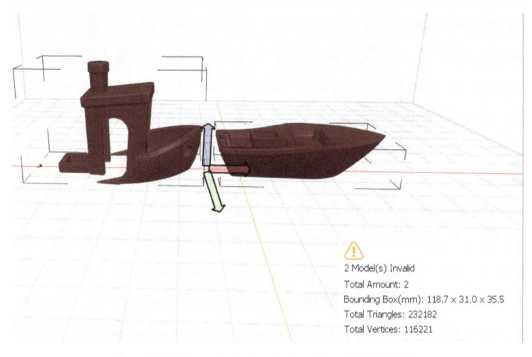

2 Model(s) Invalid
Total Amount: 2
Bounding Box(mm): 118.7 x 31.0 x 35.5
Total Triangles: 232182
Total Vertices: 116221

Figure 2.9 – The result of using the Free Cut tool in ideaMaker

5. There are now two models, and as we can see in *Figure 2.9*, they are invalid. Clicking on one, we see an error indicating non-manifold edges. To fix these errors, we need to click on the **Repair** icon at the top.

What Does Non-Manifold Mean?

We may see an error message about non-manifold edges with our objects after importing them into ideaMaker. To put it simply, non-manifold refers to shapes that cannot exist in the real world, such as a wall without a thickness. It is a good idea to fix non-manifold errors, as our 3D printer will not know how to print a part with such errors correctly.

We have successfully cut and repaired an object in ideaMaker. Now, let's look at texturing.

Applying textures in ideaMaker

One of the most exciting features in ideaMaker is the ability to add textures to an object before printing it. Adding a texture not only strengthens the object but helps in taking away the attention that layer lines have in a 3D print, as shown in the following photo:

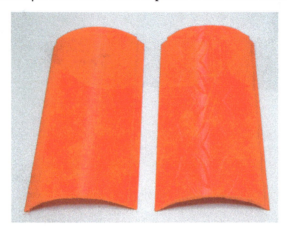

Figure 2.10 – A non-textured part versus a textured part using ideaMaker

In *Figure 2.10*, we can see the effect of applying textures. The part on the left in the photo was printed without texturing, while the part on the right had a texture applied to it in ideaMaker.

As we can see, texturing adds a dramatic effect to a part. There are many textures to choose from on the ideaMaker website. Let's try some texturing ourselves. To do so, follow these steps:

1. Load a new 3DBenchy object into ideaMaker (delete the objects from the previous section if they are still present).

2. Select the 3DBenchy object and click on the **Texture** icon at the top.

3. From the drop-down menu, select **Custom Texture**.

4. Click on the **More** drop-down menu and select **Import from ideaMaker Library**. Observe that our web browser is opened, and we are taken to the textures page of the ideaMaker website.

5. Select a pattern; in *Figure 2.10*, the **Asian Wealth** pattern was used. Observe that we are taken to a web page that is specific to the pattern.

6. Click on the **Import to ideaMaker** button. Observe that a pop-up window showing the URL for the texture pops up:

Figure 2.11 – Importing a texture from the ideaMaker website

7. Click on the **Copy** button.

8. Observe that when we return to ideaMaker, we are presented with a **Download Texture** dialog box. Click on the **Download** button to download the texture.

9. Click on the **Next** button.

10. Click on the **Yes** button to override the texture parameters.

11. Observe that our 3DBenchy object is now covered with a pattern:

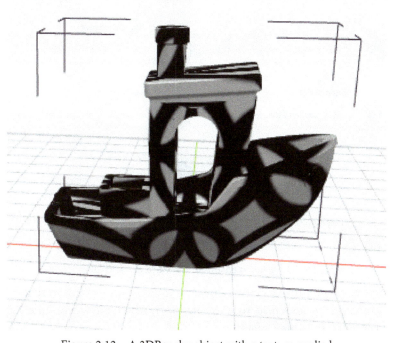

Figure 2.12 – A 3DBenchy object with a texture applied

Texturing in ideaMaker only affects the sides of an object and not the top or bottom once it is sliced. Now that we have taken a brief look at ideaMaker, let's look at Cura.

Cura

At the time of writing, Cura is the most popular 3D printer slicing software. Cura is developed by the Dutch company Ultimaker and is an open source program that is available to download for free.

Being as popular as it, Cura has an extensive large third-party plugin marketplace. Plugins for such things as OctoPrint integration and OpenSCAD file support are some of the notable additions we can add to our Cura installation. *Figure 2.13* is a screenshot of some of the plugins available in Cura:

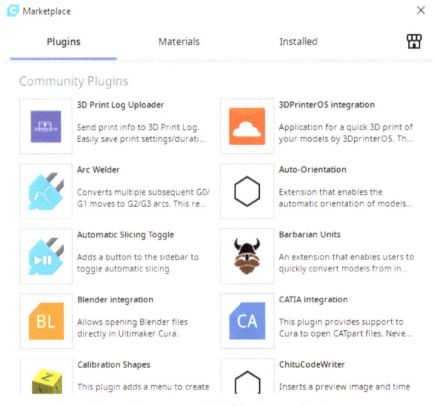

Figure 2.13 – A screenshot of the Cura marketplace

In addition to the many plugins available, Cura has many features built in that can be used for slicing our objects. One such feature is tree support, which creates unique support structures that resemble the trunk of a tree.

> **What Are Supports in Slicing?**
>
> For many objects, there are sections that are suspended from the air and require supports built from the build plate to print them. Think of an extended arm in a model of a person. Our 3D printer would not be able to print such a thing without a support structure built from the build plate. Cura comes with two different support types, normal or tree. Each support type is designed to be easily removed after printing.

Let's get some hands-on experience with tree support by applying it to an object.

Using tree support on an object in Cura

To gain a better understanding of tree support, let's load a model and slice it. To do this, follow these steps:

1. In a web browser, navigate to `https://ultimaker.com/software/ultimaker-cura` and install Cura.

2. An object with a significant overhang demonstrates the use of tree support quite well. For this example, we will use the popular Baby Yoda model from *MarVin_Miniatures* on Thingiverse. Navigate to `https://www.thingiverse.com/thing:4038181` and download the Baby Yoda model.

3. The file we are looking for is called `Baby_Yoda_v2.2.stl` and is in the `files` folder.

4. In Cura, click on **File | Open File(s)** and load the `Baby_Yoda_v2.2.stl` object:

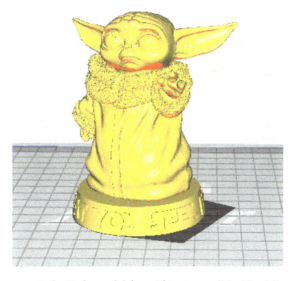

Figure 2.14 – Baby Yoda model from Thingiverse (MarVin_Miniatures)

5. To orbit in Cura, we click and hold the right mouse button while moving the mouse. If we click and hold the middle mouse button, we should be able to pan. Using these techniques, we can get an all-around view of the model.

6. When first installed, Cura defaults to the basic visibility for settings. To see all the settings, click on **Preferences | Configure Cura... | Settings | Check all**:

Figure 2.15 – Setting visibility in Cura

7. Click on **Close**.

8. Now that we have access to all the settings, it is time to add some support to our Baby Yoda model. Click on the top-right panel to get a view of the print settings.

9. Expand the **Support** selection and click on **Generate Support**:

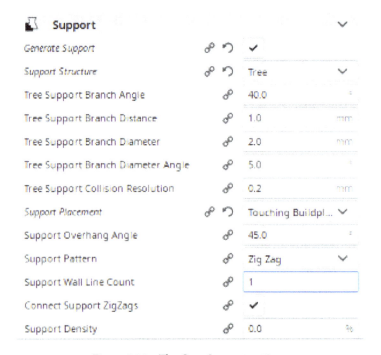

Figure 2.16 – The Cura Support settings

10. For **Support Structure**, choose **Tree** from the dropdown.

11. Close **Print settings** by clicking on **X** at the top.

12. Click on the blue **Slice** button at the bottom right-hand side of the screen.

13. Observe the **Slicing...** message at the bottom right-hand side of the screen.

14. After a short while, observe that the message box at the bottom-right has changed. Information on how long the print job will take as well as the amount of filament needed to print the object is present. A **Preview** button is present as well, which you should click.

15. Observe a newly created tree-like support structure around the Baby Yoda model:

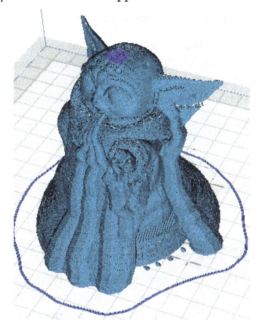

Figure 2.17 – Baby Yoda with tree support in Cura

We now have some hands-on experience with both ideaMaker and Cura. We can use this knowledge in the upcoming chapters as we slice and 3D-print objects.

Other programs we can use

Our list of slicer programs is by no means a complete list. The following is a list of other programs we can consider:

- **PrusaSlicer** is popular with Prusa-made printers. PrusaSlicer is based on Sli3r and can also be used with non-Prusa 3D printers.

- **Simplify3D** is one of the few 3D printer applications with a cost ($149 USD). Simplify3D boasts customizable supports, although that advantage has disappeared in the last few years.

- **Skeinforge** is a Python-based slicer and is considered a more complicated program to master. We can find the Skeinforge slicer engine in the Repetier-Host application.
- **Creality Slicer** comes with Creality 3D printers and is a version of Cura optimized for Creality 3D printers.

As we can see, there are many options we can choose from for slicing. Picking the right slicer may take some time.

Which FDM slicer should I choose?

Choosing the right slicing software for our projects may take some research. Slic3r has traditionally been at the cutting edge; however, other slicers are innovating quite quickly. At the time of writing, the popular program for slicing and preparing print jobs is Cura, although ideaMaker with its texturing abilities is starting to become more popular.

What it really comes down to is personal preference. For some of us, one application may be more intuitive than another. For the duration of this book, however, we will be mostly using Cura.

Before we delve into FDM slicers, let's expand our knowledge on slicers with a look at slicing programs for liquid resin printers.

Slicer programs for liquid resin 3D printers

As we will be working with FDM 3D printers throughout this book, we will be using FDM slicers such as ideaMaker and Cura. However, understanding that there is a whole different suite of slicers for 3D printers other than FDM 3D printers is good knowledge to have when we decide to venture beyond FDM 3D printing in the future.

Liquid resin 3D printing is a technology that has dramatically come down in cost to the point now where an entry-level liquid resin printer costs around $300 USD. So, what is liquid resin printing, and what are the software options available? We will start off by answering the first question.

What is liquid resin printing?

A liquid resin 3D printer is one where a series of images is projected onto a build plate through a clear bottom VAT (tank). This process is visualized in *Figure 2.18*:

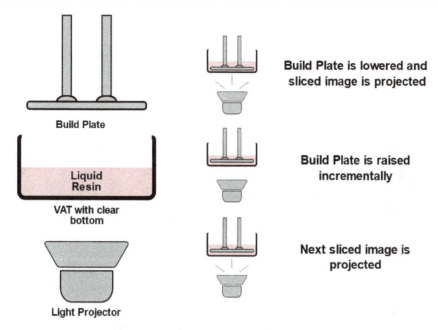

Build Plate

Liquid
Resin

VAT with clear
bottom

Light Projector

**Build Plate is lowered and
sliced image is projected**

**Build Plate is raised
incrementally**

**Next sliced image is
projected**

Figure 2.18 – A visual representation of liquid resin printing

As we can see, the constant incremental raising and lowering of the build plate in the VAT of photosensitive liquid resin creates an object on the build plate. The set layer height determines the size of the increment. So for example, if the layer height is set to 0.05 mm then the amount the build plate increments on each cycle is 0.05 mm.

> **Why Use a Liquid Resin Printer?**
>
> Liquid resin printers tend to have smaller build areas than FDM printers and require **Personal Protection Equipment** (**PPE**) when handling toxic resins. Also, a curing process is required after the print is done. Despite these challenges, the detail that a liquid resin printer provides on small prints is exceptional. This makes liquid resin printers ideal for small figurines and electronic component cases.

Unlike FMD 3D printers, liquid resin printers do not handle G-code but instead slices of images. Each image represents a layer of the object and is projected onto the build plate for a few seconds. By its very nature, liquid resin printing excels at fine details, as the resolution is not determined by the diameter of a nozzle.

Let's look at a couple of the slicing programs available for liquid resin printing.

Chitubox

Chitubox is a free 3D printer slicing software program available for Windows, macOS, and Linux. Unlike many vendor-specific slicers, Chitubox produces files that can be used on many different liquid resin printers, such as the Anycubic series of liquid resin printers.

In *Figure 2.19*, we can see a screenshot of Chitubox. A case for a Raspberry Pi Zero sits in the build plate area:

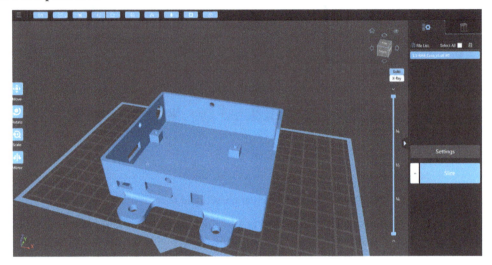

Figure 2.19 – A screenshot of Chitubox

Clicking on the **Slice** button brings us to a page where we can analyze each layer before creating a file for our liquid resin printer:

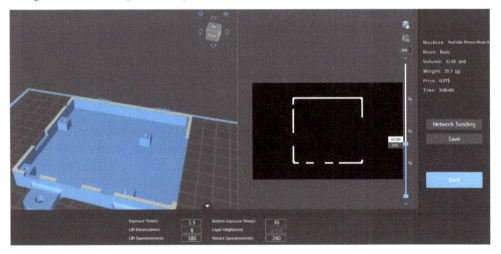

Figure 2.20 – The slicing page in Chitubox

We can also change properties such as **Exposure Time** and **Lift Distance** on this page. These properties as well as others are dependent on the resin and the printer used. In *Figure 2.20*, we can see that the print job will take 1 hour and 8 minutes, which is relatively fast.

The Anycubic Photon Workshop

Unlike Chitubox, the Anycubic Photon Workshop is designed for use by Anycubic liquid resin printers. We can see a screenshot of this program in *Figure 2.21*:

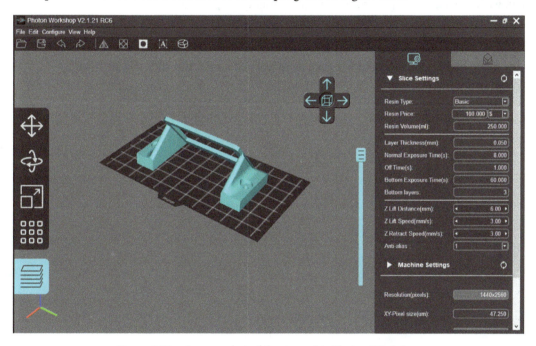

Figure 2.21 – A screenshot of the Anycubic Photon Workshop

As we can see, the slicing properties are available on the main page of the Anycubic Photon Workshop. There is also a slider that lets us view the different layers of our model. Although the Anycubic Photon Workshop does not support printers other than the Anycubic brand, the program is an excellent option for those of us with an Anycubic liquid resin printer.

Summary

In this chapter, we learned about G-code and how we can use it to control a 3D printer. Using Pronterface, we were able to control our printer through the interface as well as through executing G-code.

We were also able to create our own bed-leveling program by writing our own G-code, automating the tedious process of moving the print head around the bed.

Then, we looked at slicer programs, getting hands-on experience with ideaMaker and Cura, two enormously powerful and popular slicer programs. We finished off the chapter by taking a brief look at a couple of liquid resin slicer programs.

In the next chapter, we will take what we have learned so far and start 3D-printing physical objects.

3
Printing Our First Object

Now that we have learned a little bit about **three-dimensional** (**3D**) printers and slicers, it's time to print our first object. In this chapter, we get hands-on experience with 3D printing. The goal of this chapter is for us to find an object, slice it, and 3D print it. The knowledge gained from this chapter will help us greatly in bringing our 3D designs to life.

Before we can print, however, we need an object to print. We also need to prepare our 3D printer and slice the object into **geometric code** (**G-code**).

In this chapter, we will cover the following topics:

- Finding objects to print
- Preparing our 3D printer
- Slicing our object
- Printing our object

Technical requirements

The following resources will be required to complete the chapter:

- 3D printer—any modern **fused deposition modeling** (FDM) printer should work; however, the Creality Ender 3 V2 will be used as an example.

- A computer with Cura installed.

- The images for this chapter may be found here: `https://github.com/PacktPublishing/Simplifying-3D-Printing-with-OpenSCAD/tree/main/Chapter3`.

Finding objects to print

Printing objects that are of no use to anyone else but us is the ultimate use of a 3D printer. For most of us, it hardly seems worth it to spend hours printing something such as an inexpensive hook for our coat that we could just buy at a store. The exception would be for remote outposts where getting such common items is a challenge.

> **3D Printing in Space**
>
> A 3D printer made by an American company **Made In Space** was first used aboard the **International Space Station** (ISS) in 2014 to print a ratchet. The 3D printer, named **Additive Manufacturing Facility** (AMF), has a build volume of 14 cm by 10 cm by 10 cm. Designed to print common filament materials such as **Acrylonitrile Butadiene Styrene** (ABS), the goal of the company is to have the printer use materials made from moon dust and Martian soil.

In this section, we will explore the various avenues where we can find objects to print. We will start with a look at 3D file formats.

Understanding 3D object file formats

To 3D print a file from our computer we need the file to contain 3D information such as geometry, texture, and color. There are many 3D object file formats; however, only a select number can be used for 3D printing.

The following are file formats we can use to 3D print with.

Stereolithography (STL)

Stereolithography (**STL**) files trace their origins to the 1980s when the format was used for 3D printers developed by the company **3D Systems**. STL files encode the surface of an object using a triangular mesh. Using more triangles in the mesh makes for higher-resolution parts but larger file sizes. STL files can be difficult to modify but may be imported into programs such as Blender, Fusion 360, and OpenSCAD for use with other shapes. For example, we can import an STL file of a Raspberry Pi into our **computer-aided design** (**CAD**) program to use as a model for making a Raspberry Pi case.

STL files may be viewed prior to printing so that we can confirm the object is what we want to print. In Windows, we can use the **3D Viewer** program to view our STL files. Holding down the left mouse button in 3D Viewer allows us to use our mouse to orbit in all directions around our object. Holding down the right mouse button gives us pan control. 3D Viewer comes pre-installed in the latest version of Windows.

In the following screenshot from 3D Viewer, we can see an STL file representing a model rocket nose cone:

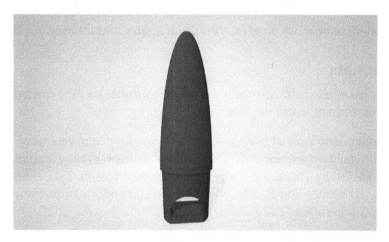

Figure 3.1 – Model rocket nose cone STL file

This nose cone was designed in OpenSCAD and exported with the .stl file format used for STL files.

STL files contain only the geometry information of a 3D object. Thus, the attributes we mentioned—such as textures or color—are not represented in the file. Despite this limitation, STL files continue to be the most popular file format for 3D printing.

With macOS, we can view STL files using the built-in program, **Preview**.

OBJ (Wavefront)

The **OBJ** file format was developed by the company **Wavefront Technologies** and its files are sometimes referred to as **Wavefront Objects**. OBJ files differ from STL files in that they store colors and textures, making OBJ files more complex than STL files.

By supporting color, OBJ files stand to gain in popularity when multi-color 3D printers become more commonplace.

SCAD (OpenSCAD)

The **Solid CAD** (**SCAD**) (`.scad`) file format is native to the OpenSCAD CAD program. SCAD files are text-based code formatted files. We will start our exploration of OpenSCAD and SCAD files in *Chapter 4, Getting Started with OpenSCAD*.

Although not thought of as a file format in which to 3D print objects, we can import SCAD files directly into Cura by utilizing the **OpenSCAD integration** plugin from the Cura Marketplace. Using this plugin simplifies our workflow as we no longer need to create an STL file from our OpenSCAD design to import into our slicer.

This will make more sense as we progress through designing and printing our own objects.

Other file formats

STL, OBJ, and SCAD files are by no means the only file formats we 3D print from. Here are some other file formats worth mentioning:

- **3D Manufacturing Format (3MF)**—With STL files limited to geometry, there is a push for a standard offering more data. 3MF is an answer to the limitations of STL files. Backed by software industry heavyweight Microsoft, 3MF aims to create a seamless 3D printer workflow by offering 3D texture information, thumbnail images, and color, along with geometry data.

- **Additive Manufacturing Format (AMF)**—Competing with 3MF is the AMF format, which was initially dubbed as STL 2.0. Along with geometry data, AMF files contain information for texture, color, and metadata (name, author, company).

- **Image files** (.png, .bmp, .jpg)—Common image formats such as .png, .jpg, and .bmp are not 3D object files but may be loaded into our slicer programs to create a raised picture 3D print. Cura and ideaMaker can both do this.

Now that we have an understanding of the file types we use for 3D printing, let's turn our attention to where we can find objects to print.

Downloading 3D objects

Before we spend time designing an object we need, it is a good practice to see if that object has already been designed and is available to be downloaded. An example of where we might want to do this is for upgrades to our 3D printer. For popular printers, such as the Ender 3 V2, there are many downloadable upgrades available.

Detailed next are resources where we can find 3D printable objects.

Thingiverse

Thingiverse (`www.thingiverse.com`) was started in 2008 as a resource for user-uploaded 3D designs by **MakerBot Industries**. Since then, it has exploded in popularity, with over 2 million 3D models uploaded as of this writing.

Thingiverse offers 3D models in various formats such as STL, OBJ, and SCAD. Many functional parts can be found on Thingiverse, and there is an option to tip the designer. An analytics dashboard is available for uploaders to view statistics as illustrated in the following screenshot:

Figure 3.2 – Analytics dashboard in Thingiverse

Being as popular as it is, Thingiverse has become the de facto repository for the maker community. Manufacturers can use Thingiverse designs as inspiration for product improvements. An example of this is the Thingiverse tool drawer upgrade for the Ender 3 that found its way onto the Ender 3 V2.

YouMagine

Associated with 3D printer manufacturer **Ultimaker**, **YouMagine** (`www.youmagine.com`) aims to be a platform for open source creation. YouMagine enforces a strict takedown policy of infringement of original designs. The top menu on the YouMagine website features **Designs**, **Collections**, and **Blog** tabs.

Under the **Designs** tab, we can browse through thousands of open source models for download. A category filter defaults to **All Categories**, but by clicking on the down arrow, the filter exposes many categories to choose from. At the time of writing, a **COVID19** category exists with **154** designs. These designs consist of various **personal protective equipment** (**PPE**) models to be 3D printed and used in the fight against COVID-19.

> **Makers to the Rescue**
>
> As impressive as the global supply chain is, there are times when it is too big or too slow to meet immediate local demand, such as when turnaround times from design to production must be measured in days or even hours, as with the shortage of PPE during the COVID-19 pandemic. In the early days of the pandemic when PPE was in short supply, the maker community mobilized rapidly to 3D print PPE for workers on the front line.

The **Collections** tab organizes 3D designs into collections, and the **Blog** tab provides a link to informative articles on 3D printing and the YouMagine environment itself.

GrabCAD

GrabCAD (`www.grabcad.com`) is an online community of designers, engineers, hobbyists, and educators sharing CAD designs and 3D models. GrabCAD was founded in 2009 in Estonia but has since moved to the **United States** (**US**).

Using the library in GrabCAD, we can download CAD design files that can be edited to suit our needs. GrabCAD is geared more toward CAD design than 3D printing, as many models would require a **computer numerical control** (**CNC**) machine over a 3D printer.

Now that we know of a few resources to acquire 3D models to print, let's turn our attention to downloading test models for calibration.

Calibration objects for our 3D printer

Ensuring that our printer is calibrated correctly is an important step for creating high-quality 3D prints. This is especially important when making functional parts that must fit with other parts. To calibrate our 3D printer, we print out a model with defined measurements in its design and then physically check those measurements with calipers. We can see three such calibration models in the following image:

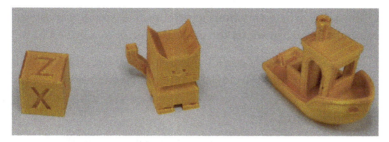

Figure 3.3 – Calibration models

From left to right are the XYZ 20mm Calibration Cube, Cali Cat, and the #3DBenchy. All three calibration models can be found on Thingiverse.

Let's take a deeper look at each model.

The Calibration Cube

The Calibration Cube can be found on Thingiverse here: `https://www.thingiverse.com/thing:1278865`. As we can see in *Figure 3.3*, the Calibration Cube is a simple XYZ cube. When printed, the Calibration Cube is measured along each axis, as shown in the following image. In this example, we see that the measurements are close:

Z axis (20.19mm)

X axis (20.13mm)

Y axis (20.12mm)

Figure 3.4 – Calibration Cube measurement results

Please note that the *z* axis is where variations in print results would be most noticeable. This is due to how close to the bed we set the print head prior to a print job.

The Calibration Cube prints relatively quickly due to its small size. Having the axes marked on the cube makes it easy for us to determine which one needs adjustment.

Cali Cat – The Calibration Cat

For a more interesting calibration model than the Calibration Cube, there is the **Cali Cat**. The Cali Cat measures 35 mm in the *z* direction or from the top of its ears to the bottom.

The head is 20 mm in the *x* direction and 20 mm in the *y* direction. The tail extends from the body at exactly 45 degrees.

An illustration of the Cali Cat is provided in the following image:

Figure 3.5 – Measuring z height on the Cali Cat

The Cali Cat can be downloaded from Thingiverse here:

```
https://www.thingiverse.com/thing:1545913
```

The #3DBenchy

Anyone who has spent some time looking at 3D printer videos on YouTube has no doubt come across the **#3DBenchy**. Disguised as a cartoonish tug boat, the #3DBenchy is a torture test for our 3D printer and slicer settings. We will be using the #3DBenchy in the remaining sections of this chapter to test our slicer settings.

The #3DBenchy can also be used to calibrate our 3D printer, as there are set dimensional properties throughout the model. In the following image, we can see the measurement of the bridge roof length. Our value of **22.92** mm is just shy of the designed value of 23 mm:

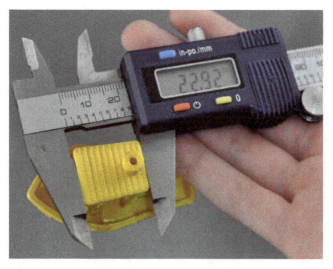

Figure 3.6 – Measuring bridge roof length on the #3DBenchy

The #3DBenchy can be found on Thingiverse here: `https://www.thingiverse.com/thing:763622`. There is also a dedicated website located at `http://www.3dbenchy.com`. The website includes an extensive analysis section of measurements designed in the model.

Now that we have an idea of where we can find 3D objects to print, including calibration models, let's now prepare our printer and print jobs and start 3D printing.

Preparing our 3D printer

It's time to get our printer ready and start printing. For the remainder of this chapter, we will focus on printing out the #3DBenchy model. We will first prepare our printer before slicing our model into G-code. We will be using **polylactic acid** (**PLA**) to print our model.

The following steps will outline using the Ender 3 V2; however, these steps can be used with other modern 3D printers.

We will start off by leveling our bed. Using the G-code we wrote in *Chapter 2, What Are Slicer Programs?*, run the bed-leveling program and proceed as follows:

1. Before we install the filament, it's a good idea to place the print head above the print bed so that we can verify that the filament has been loaded correctly. To do this, click on **Prepare** | **Move** | **Move Z** and set the value to **20**.

2. We will now load our PLA filament into our 3D printer. To do this, we have to raise the temperature of the nozzle. Using the control knob on the control panel, navigate to **Control** | **Temperature** | **Nozzle** (or **Hotend** if the firmware has been updated).

3. Set the temperature to **200**°C and click the control knob to accept.

4. Observe that the temperature starts rising.

5. At the back of the printer at the extruder motor, lift off the blue dial and unscrew the filament tube coupler.

6. Slide the PLA into the hole at the side and through the extruder motor. We may have to press the extruder motor lever to get the filament through, as illustrated in the following picture:

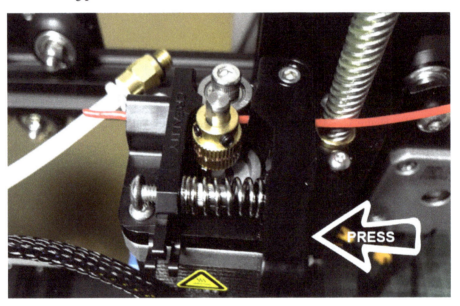

Figure 3.7 – Loading filament into the extruder

7. It may take a few tries to thread the filament. Once it is through, screw the filament tube coupler back onto the extruder motor assembly and put the blue dial back in place.

8. Once the nozzle temperature reaches **200**°C, we can start to push the filament through the filament tube. Turn the blue dial counterclockwise to move the filament through the tube. Keep going until the filament starts to extrude out of the nozzle, as illustrated in the following picture:

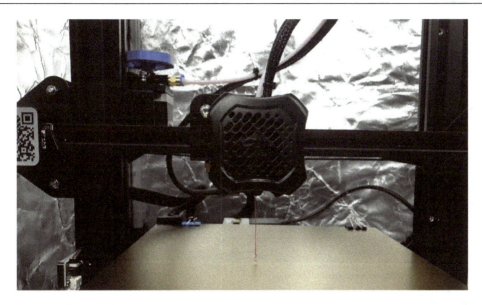

Figure 3.8 – Filament extruding through the nozzle

9. To avoid the possibility of a filament jam before we need to print, we should bring the nozzle temperature back down. Using the display panel, click on **Prepare | Cooldown** or use **Control | Temperature | Nozzle** and set the temperature to **0**.

Our printer is ready to print, and we are now ready to create a print job using a slicer program.

Slicing our object

We will slice the #3DBenchy model using Cura. In the process, we will get to know some of the more common settings in Cura. If you haven't already done so, refer to the *Finding objects to print* section of this chapter, and the *Common FDM slicer programs* section of *Chapter 2, What Are Slicer Programs?* to download the #3DBenchy and install Cura, respectively.

With Cura installed, let's dive into the settings.

Setting up the profile

To slice an object for our printer, Cura needs to know which 3D printer we are using. We configure this either when first installing Cura or later using the **Add Printer** button. We will be going through the steps to add a printer to an existing Cura installation. To do this, we proceed as follows:

1. To access the **Add Printer** dialog, click on **Settings | Printer | Add Printer...** from the main menu, which will take you to the following screen:

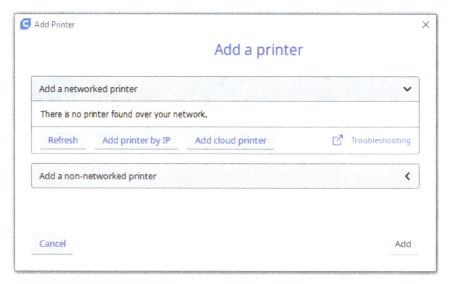

Figure 3.9 – Setting up a 3D printer in Cura

2. Select **Add a non-networked printer** and then **Creality Ender Pro** under the **Creality3D** heading. Change the **Printer name** setting to **Creality Ender-3 V2**.

3. Click on the **Add** button.

4. Click on **Next** again to select the default settings.

5. Observe in the following screenshot that our printer is now set up with a **Generic PLA 0.4mm Nozzle** profile:

Figure 3.10 – 3D printer profile in Cura

We will be using the **Generic PLA 0.4mm Nozzle** profile as a starting point and will modify a few settings to create our own material profile.

Now that we have our printer set up in Cura, let's load our model and start to configure the slicing options.

Loading our model

We will use the #3DBenchy model, as described in the *Finding objects to print* section. To load our model, we do the following:

1. From the main menu, click on **File | Open File(s)....**

2. Locate the 3DBenchy.stl file and click on **Open**.

3. Observe that the #3DBenchy model is loaded onto the middle of the build plate in Cura.

   Before we can change any slicer settings in Cura, we must have access to all of them. From the main menu, click on **Settings | Configure setting visibility**.

4. Click on the **Check all** box under the **Setting Visibility** heading.

Now that we have prepared our printer and loaded our model, let's start changing some settings in Cura.

Quality settings

Settings made in the **Quality** section are among the most important settings in determining the success or failure of our 3D prints. To access the **Quality** settings, click on the section just below the **Marketplace** button and then on the side arrow to open it up so that your screen now looks like this:

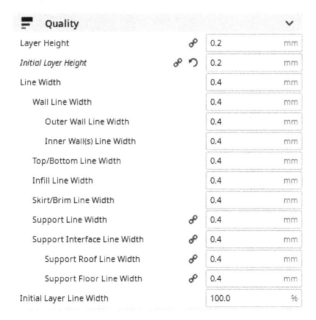

Figure 3.11 – Cura Quality settings

We will now look at and change some of the **Quality** settings, as follows:

- **Layer Height**—Layer height is the distance that the *z* axis travels for each layer. Lower values result in better quality but longer print times. For good measure, the layer height should not be more than 80% of the nozzle diameter—ideally, around 50% for maximum layer adhesion. Type in **0.2** for the **Layer Height** setting.

> **Ender 3 Magic Number**
>
> Due to the *z*-axis lead screw and steps of the *z*-axis stepper motor, the Ender 3 has a "magic number". This is a number where if it is divisible by the layer height, it will produce prints with improved quality. In the case of the Ender 3, the magic number is 0.04. Thus, layer heights in increments of 0.04 (0.08 mm, 0.12 mm, 0.16 mm, 0.2 mm) will produce better-quality prints than other layer heights (0.1 mm, 0.13 mm). For more information on this, check out CHEP's video at `https://www.youtube.com/watch?v=WIkT8asT90A`.

- **Initial Layer Height**—This setting determines the initial height of the print head off the bed. The actual height will be the value stored here plus the height set when the bed is leveled. Generally, this value is set to 0.2 mm or 0.3 mm. Some prefer 0.3 mm as it extrudes more material onto the build plate, while others prefer 0.2 mm as it sticks to the bed better. Type in **0.2** for the **Initial Layer Height** value.

- **Line Width**—This sets the width of the line produced from the nozzle. Generally, this value is the same as the nozzle diameter but can be adjusted if it appears our extruder is underextruding or overextruding. If this is the case with the extruder, then other parameters should be looked at first (such as steps per mm for the extruder) before setting **Line Width** to anything but the nozzle diameter. As our nozzle has a 0.4 mm hole diameter, set all **Line Width** values to **0.4**.

Now, let's take a look at the **Infill** settings.

Infill settings

One of the more fascinating aspects of FDM 3D printing is the ability to change how dense an object is by setting the infill percentage. For example, suppose we were making wings for a model aircraft. Using a more traditional method such as injection molding, we would probably make the part a solid part.

In the following screenshot, we can see a cross-section of the wing (an airfoil) for our model plane as an injection-molded solid part (left) and as a 3D-printed part with a 20% infill (right):

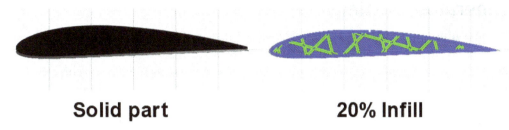

Figure 3.12 – Solid airfoil versus one with 20% infill

With a solid part, more material is used, and thus the wing is heavier. Using a 20% infill we may be able to achieve the necessary rigidity and not only save on material but make the wing lighter.

We will look at two values in the **Infill** settings—**Infill Density** and **Infill Line Multiplier**. To adjust these values, we do the following:

1. Click on the side arrow in the **Infill** row to view the **Infill** settings.
2. In the **Infill Density** box, put in the value **20**.

The **Infill Line Multiplier** setting increases the number of lines the slicer uses when creating an infill. Increasing the **Infill Line Multiplier** value allows us to reduce the amount of infill to save on filament and keep the strength of a higher infill percentage. In the following screenshot, we can see the difference between an **Infill Line Multiplier** setting of 1 versus 2:

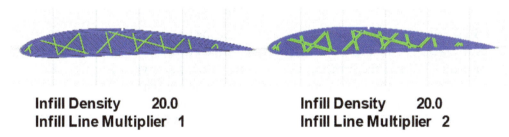

Figure 3.13 – Infill Line Multiplier from Cura

For our example, we will set the **Infill Line Multiplier** value to **2**. Put a value of 2 into the **Infill Line Multiplier** box.

There is a balance between decreasing the **Infill Density** value and increasing the **Infill Line Multiplier** value to reduce filament usage. Ideally, we would like to maintain the strength of using a higher **Infill Density** value.

Now that we have looked at the **Infill** settings, let's move on to temperature.

Temperature settings

It may be confusing that there is no temperature settings section in the Cura slicer. Instead, temperature has been placed under the **Material** settings section. Temperature is one of the most important settings to get right. It is also a setting that can be changed during the print job.

Why Change the Temperature During a Print Job?

Ideally, we will set the temperature correctly in the slicer and not have to worry about it. However, there are times when we need to adjust the temperature during a print. For example, we might hear a clicking sound coming from the nozzle if the nozzle temperature is not high enough to allow the filament to flow through at the set rate. Raising the temperature a few degrees may be all we need to do to fix such an issue. As well, we might notice that the filament is giving the part a melted look (*see Figure 3.15*). This may be caused by printing a smaller object than the one used when the profile was set up. Lowering the nozzle temperature should fix this issue.

To set the nozzle temperature for our example print, we do the following:

1. Click on the side arrow in the **Material** row to view the **Material** settings.

2. For **Printing Temperature** and **Final Printing Temperature**, enter the value **200**.

3. The **Printing Temperature Initial Layer** setting is used to set the nozzle temperature for the first layer of our print job. Setting a higher temperature than the **Printing Temperature** value helps in adhesion for the first layer. Enter **205** for this value.

4. **Initial Printing Temperature** is the value where the print job will start. Having this value lower than the **Printing Temperature Initial Layer** value allows us to kick off the print job before the **Printing Temperature Initial Layer** value has been reached. This is useful in situations where we want our print job to start earlier as it is taking too long for the printer to reach the **Initial Printing Temperature** value. We should not have this issue with the Ender 3 V2. Set this value to the same as for **Initial Printing Temperature—205**.

Now that we have the nozzle temperature set, it's time to set the bed temperature. For certain materials such as ABS, having a high bed temperature is necessary to ensure that the filament sticks to the bed. PLA is far more forgiving in this respect.

It is possible to print PLA to a non-heated bed, and in fact, many of the early 3D printers did not come with heated beds. For our example, we want to take advantage of the stickiness of PLA when it is applied to a heated bed.

To set the bed temperature for our print, we do the following:

1. For **Build Plate Temperature**, put in a value of **60**.

2. **Build Plate Temperature** allows us to set a different build temperature for the first layer. When using a higher temperature than the **Build Plate Temperature Initial Layer** value, this helps in creating adhesion to the bed. We will set this to the same value as for **Build Plate Temperature**. Put in a value of **60** for **Build Plate Temperature Initial Layer**.

Now that we have the nozzle and bed temperatures set, let's look at the cooling settings for the print job.

Cooling settings

On the right side of the Ender 3 V2 extruder is the **part cooling fan** (*Figure 1.5* from *Chapter 1, Getting Started With 3D Printing*). The aptly named part cools the filament after it has been extruded. The strength of the fan—and thus its cooling effect—is adjustable. In the following screenshot, we see how the cooling fan works. As the filament settles on to the layer, the fan cools the layer:

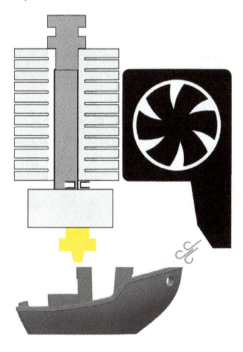

Figure 3.14 – Part cooling

For some filaments such as ABS, part cooling (especially excessive part cooling) works against layer adhesion and is generally not used very much. PLA, on the other hand, requires part cooling, with quality suffering without it. We can see an extreme case of not cooling down a PLA part in the following picture:

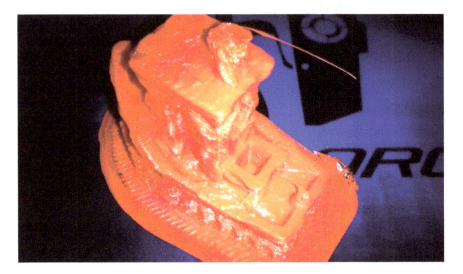

Figure 3.15 – PLA part without part cooling

It is a good idea to limit part cooling on the first few layers as it affects bed adhesion. For our example, we will limit part cooling for the first four layers. To do this, proceed as follows:

1. Click on the side arrow in the **Cooling** row to view the **Cooling** settings.

2. Click on the **Enable Print Cooling** checkbox if it is not already checked.

3. For **Fan Speed**, put in the value **100**.

4. For **Initial Fan Speed**, put in the value **0**.

5. For **Regular Fan Speed at Layer**, put in the value **4**.

What we have done is limited part cooling for the first four layers, turning it off completely for the first layer. This, along with the higher initial nozzle temperature, will assist in having the filament stick to the bed for the first layer. The cooling effect from the part cooling fan will increase every layer and will be at 100% at layer four.

Let's move on to the last setting we will modify—the adhesion to the build plate.

Build Plate Adhesion settings

The **Build Plate Adhesion** settings determine how our object adheres to the build plate. With 3D printing, getting the first layer to stick to the build plate is the most important step in the whole process.

There are four different settings for **Build Plate Adhesion Type**: **None**, **Brim**, **Skirt**, and **Raft**, as illustrated in the following screenshot:

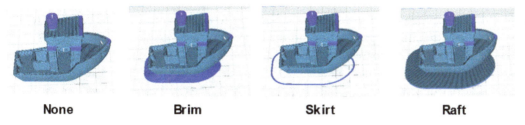

None   **Brim**   **Skirt**   **Raft**

Figure 3.16 – Build Plate Adhesion types in Cura

Let's take a look at each type now.

None

By setting the **Build Adhesion Plate Type** to **None**, our print head will immediately start printing our object once the cleaning function has run. This value is desirable in situations where space is tight, as the other three adhesion types require extra space on the build plate.

> **Nozzle Cleaning Function**
>
> A nozzle cleaning job is run just before every print job. This is made up of pre-written G-code that our slicer inserts into our G-code file. For our Ender 3 V2, there are lines extruded on the left side from the front to the back and from the back to the front. This extrusion helps clean out the nozzle before the object is printed and happens at the beginning of each print job.

Brim

Brim takes its name from the brim of a hat. It adds width to the base of our print on the first layer and can be easily removed after the print job is done. Brims are especially useful for prints that have wide bases and help to protect against the bottom edges of our print curling up.

Skirt

Skirt is probably the most popular **Build Plate Adhesion Type**. It basically adds extra lines to the nozzle cleaning function around the print and allows us to level the bed in real time before our object is printed. We do use a skirt for our example print in the upcoming section, *Printing our object*.

Raft

Before the widespread use of glass beds, the best way to ensure a flat build surface was to have it made at the time of printing. A raft made up of a few layers of filament provides excellent adhesion for our objects as the objects are printed on top of the very same material they are made of. The **Raft Air Gap** setting determines how easily the raft can be removed from our print after the print job has finished. The right combination of nozzle, bed temperature, and raft air gap may take some experimentation to get it right. As useful as they are, rafts have fallen out of favor due to the extra filament and print time they require.

For our example, we will use the **Skirt** adhesion type. To do this, we proceed as follows:

1. Click on the side arrow in the **Build Plate Adhesion** row to view the **Build Plate Adhesion** settings.

2. Select **Skirt** from the **Build Plate Adhesion Type** drop-down menu.

3. Adding extra lines to a skirt gives us more time to live adjust our print bed. For **Skirt Line Count**, put in the value **10**.

4. We will leave the other settings at their default values. As this is the last setting, we will modify it as it's time to save our profile. From the **Profile** drop-down box at the top, select the down arrow.

5. Select **Create profile from current settings/overrides....**

6. Type in a name in the **Create Profile** dialog box.

7. Click **OK** to save.

8. Click **Close** to close the **Preferences** box.

With the settings done, it's now time to slice our object.

Slicing our object

To slice our object and create and save the G-code, we need to do the following:

1. Insert a microSD card into the computer using an appropriate adapter.

2. In Cura, click on the blue **Slice** button on the bottom right-hand side.

3. Click on the **Save to Removable Drive** button.

4. Eject the microSD card using the appropriate operating system function or the dialog popup in Cura.

Now that we have learned a bit about slicer settings, the time has come to bear the fruit of all our efforts up to this point and print out our object.

Printing our object

With our print bed leveled, filament loaded, and object sliced, it's now time to run our print job. For our example, we will be using the standard glass bed that comes with the Ender 3 V2.

Applying a glue stick to the bed

Although we could probably print directly to the glass bed, it is a good idea to apply a glue stick to assist in first-layer adhesion.

Applying a glue stick in one direction followed by the other, then waiting a few minutes between the first and second application will help create a sticky surface for our PLA, as illustrated in the following picture:

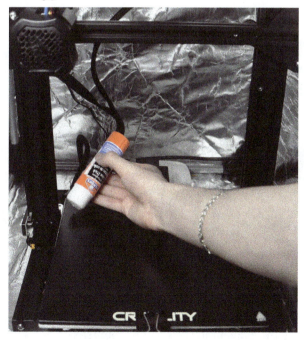

Figure 3.17 – Applying a glue stick to our print bed

Now that our print bed is ready, let's start the print job.

Running our print job

We run the print job from the microSD card. While observing the printing of the skirt, we can determine any action that needs to be taken to level the bed further. To do this, we proceed as follows:

1. Load the microSD card into the 3D printer. For our Ender 3 V2, the slot is located on the front left side.

2. Click on **Print**, scroll to the **3DBenchy** file, and click to select. Observe that the print job starts.

3. As we have set the **Skirt Line Count** value to 10, we have some time for adjustments. If the skirt is not sticking to the bed at all, we adjust the z offset. To do this, click on **Tune**, scroll down to **Z-Offset**, click and enter a negative value (**-0.2**, for example), and then click again to accept.

4. If it appears that only one side is not sticking, then the bed needs adjustment. Gently adjusting the appropriate leveling wheel should fix the issue. We must be careful when doing this as the bed is hot and we do not want to stop the bed from moving, as it will cause a shift in our print.

Ideally, our skirt will stick evenly to the bed, as shown in the following image:

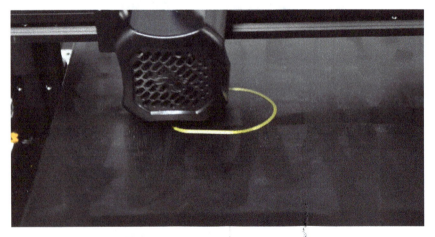

Figure 3.18 – Skirt sticking to the bed

By adjusting the z offset or adjusting our bed, we run the risk of negatively affecting the accuracy of the z height of our object. If the z height is important for our print, then we could consider adjusting the z height in the slicer to compensate by using scaling or by choosing a raft for our **Build Plate Adhesion Type**.

> **Adjusting Z Size in Cura**
>
> In Cura, looking at the buttons on the left-hand side, the second from the top
> is **Scale**. This is used to resize objects. By turning off **Uniform Scaling**, we can
> adjust the *z* height of our objects prior to slicing to allow for any inaccuracies.

After our print is done, we can inspect our object for printer accuracy and print settings.
The correct dimensions for the #3DBenchy are well documented here: `http://`
`www.3dbenchy.com/dimensions/`.

Calibrating our printer

If the measured values from our printed #3DBenchy are significantly off, then we need to
adjust the **steps per mm** value for the affected axis. For the Ender 3 V2, we can access this
value from **Control | Motion | Steps/mm**.

To correct the affected axis, calculate and put in a new value for steps per mm by using the
following formula:

```
NewStepsPerMM = (CorrectValue * CurrentStepsPerMM) /
MeasuredValue)
```

For example, in *Figure 3.6*, we can see a measured #3DBenchy Bridge roof length of 22.92
mm. This measurement represents the *x* axis. To calculate `NewStepsPerMM` for *x*, use the
following formula:

```
NewStepsPerMM = (23 * 80) / 22.92)= 80.28
```

The `CurrentStepsPerMM` defaults to 80 on the Ender 3 V2. As we can see, the new
value is very close to the existing steps per mm value.

Inspecting #3DBenchy for print quality

The humble #3DBenchy offers us quite a lot of insight into the accuracy of our slicer
settings. For our purposes, we will look at the bottom of our print and determine what
we need to do to achieve a great first layer.

> **Detailed #3DBenchy Analysis**
>
> For an excellent analysis on the #3DBenchy and how it ties in with slicer
> settings, check out YouTuber Michael Law's video:
>
> `https://www.youtube.com/watch?v=t_7EMnQ6Rlc`

In the following picture, we can see two separate #3DBenchy prints:

Figure 3.19 – #3DBenchy first layer test

On the black print on the left, we can clearly see the test text. This indicates that our first layer was successful. For the gold print on the right, the test text is washed out, indicating that our print head was too close to the bed. It could also mean that the bed or nozzle temperature was too high.

As already stated, getting the first layer right in a 3D print job is the most challenging part of 3D printing as the rest of the print job relies on it. Tweaking slicer settings and build plate surfaces goes a long way toward getting great first layers.

We have now learned how to print and inspect an object.

Summary

We started this chapter by looking at 3D object types such as STL and OBJ. We then investigated where we could find objects to print. With the incredible amount of 3D designs available, it serves us well to know where to find designs that suit our needs before venturing into creating our own.

We then looked at some common slicer settings and modified them. Using a default slicing profile to work from saves us from having to learn all the settings available. We explored layer heights, infill settings, and temperature and cooling settings as these are some of the more important settings to understand to create high-quality 3D prints.

Using the standard Creality Ender 3 V2 glass build plate, we printed a #3DBenchy test print and did some analysis on it to determine the accuracy of our first layer, by far the most important layer for a successful 3D print.

In the next chapter, we will start learning OpenSCAD as we shift the focus to designing our own objects.

Part 2: Learning OpenSCAD

Although 3D printing objects designed by others is certainly worthwhile, having the ability to create your own 3D designs expands what you can do greatly. Physical objects do not need to stay in your imagination when you are armed with the tools of 3D design. OpenSCAD is a free, open source 3D design platform designed with the programmer in mind. In this part, we will explore OpenSCAD. We will start off designing with simple shapes and work our way to more complex designs.

In this part, we cover the following chapters:

- *Chapter 4, Getting Started with OpenSCAD*
- *Chapter 5, Using Advanced Operations of OpenSCAD*
- *Chapter 6, Exploring Common OpenSCAD Libraries*

4
Getting Started with OpenSCAD

Although 3D printing objects designed by others is certainly worthwhile, having the ability to create our own 3D designs expands what we can do greatly with our 3D printer. Physical objects do not need to stay in our imagination when we are armed with the tools of 3D design.

OpenSCAD is a free, open source 3D design platform designed with the programmer in mind. While having prior programming knowledge is an asset in learning OpenSCAD, it certainly is not a requirement.

We will start off designing with simple shapes and work our way to more complex designs. We will use our knowledge to create a customized hook for a PVC pipe.

> **Important Note**
>
> Our examples will become increasingly more difficult in this chapter. Care must be taken that the correct code is entered to avoid syntax errors that may require a significant time to fix.

In this chapter, we will cover the following topics:

- Introducing OpenSCAD
- Exploring other CAD programs

- Learning OpenSCAD GUI and basic commands
- Learning OpenSCAD Boolean and transformation operations

Technical requirements

The following will be required to complete the chapter:

- Any late model Windows, macOS, or Linux computer that can install OpenSCAD.
- The code and images for this chapter may be found here: `https://github.com/PacktPublishing/Simplifying-3D-Printing-with-OpenSCAD/tree/main/Chapter4`.

Introducing OpenSCAD

OpenSCAD is often referred to as a programmer's 3D design tool. Unlike many CAD environments out there, OpenSCAD designs are created by writing C-like code in an editor. Shapes created by commands may be added and subtracted from one another to create new objects. Re-usable functions and libraries may be easily written.

As an example, in *Figure 4.1*, the code `cube([10,10,10], center=true);` creates the cube we see on the right:

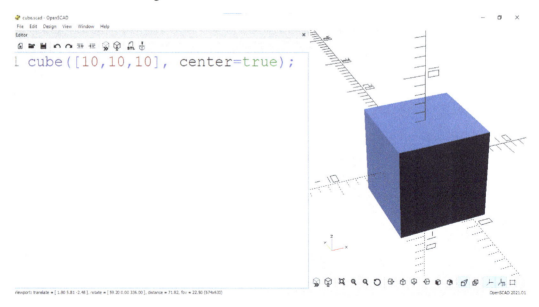

Figure 4.1 – Cube made in OpenSCAD

Objects for use in the real world may be easily designed using OpenSCAD code and exported to be printed on a 3D printer. *Figure 4.2* shows a console tray for a 2012 Toyota Prius designed in OpenSCAD. This design was created using only a few lines of code:

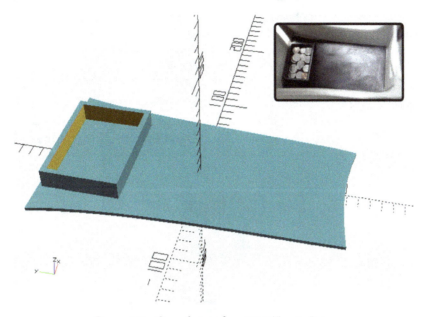

Figure 4.2 – Console tray for a 2012 Toyota Prius

By virtue of being a programming environment, OpenSCAD is highly parameterizable. Parts created may be scaled or modified easily by changing parameters. This is demonstrated in *Figure 4.3*.

We can see two PVC pipe hooks. These hooks are designed to screw into a PVC pipe and used for hanging such things as tools or jackets. The hook on the left is designed for a PVC pipe that has a diameter of 42.5 mm, and the hook on the right is designed for a PVC pipe with a diameter of 32.5 mm, as shown here:

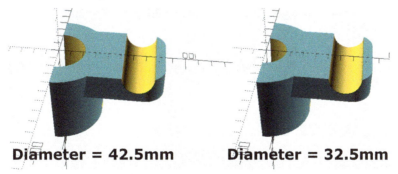

Figure 4.3 – Hooks for different size poles

The **Diameter** value is stored as a number in the program and is quite easily modified to suit the diameter of the PVC pipe used.

PVC Pipe and 3D Designs

When designing parts that require long cylinder shapes, it is a good idea to incorporate PVC pipes in the design. PVC pipes are relatively inexpensive and are much stronger than a vertically printed 3D cylinder.

Being a programming environment, there are outside libraries that may be used to help simplify the design. To generate the M3 15-mm bolt shown in *Figure 4.4*, we use the **BOSL** library available for download from the OpenSCAD website (`http://openscad.org/libraries.html`):

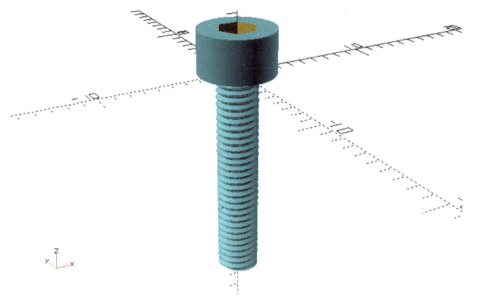

Figure 4.4 – Bolt generated in OpenSCAD using the BOSL library

With the BOSL library, the M3 15-mm bolt is generated with just one line of code.

Now that we have a general idea of what OpenSCAD is, let's see how it stacks up against other CAD software.

Exploring other CAD programs

Choosing the right CAD software for a project can be a difficult task. The learning curve for any CAD environment is a steep one. To maximize the return on our time, we must take care to choose the right CAD software for our intended application.

Let's look at some of the alternatives we may use.

Fusion 360

Fusion 360, by the American company AutoDesk, combines CAD, CAM, and PCB design in one package. Usage is based on a subscription model, although a free version is available for personal use. Designs in Fusion 360 generally start from a sketch. Constraints and dimensions applied to a sketch may be modified later in the design as Fusion 360 maintains a design history.

In *Figure 4.5*, we can see a table for a vacuum form machine that was designed in Fusion 360:

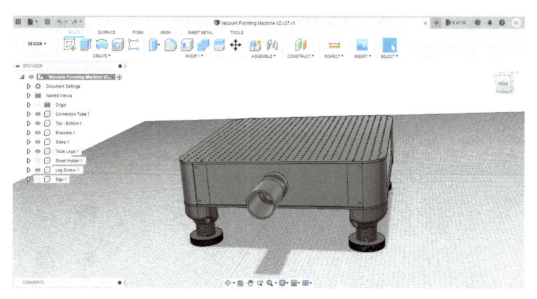

Figure 4.5 – Table for a vacuum forming machine

This design was built from extruded sketches. The filet functionality used in this design gives it a distinctive look.

What Are Filets?

A filet is a rounding of an inside or outside edge in a design. Filets may be applied for mechanical reasons such as stress concentration or simply for design aesthetics.

Fusion 360 operates as a hybrid desktop/online model with the program installed on a local computer and projects stored in the cloud. You may access your designs from any computer logged in to your account.

The functionality of Fusion 360 is vast. 3D printing from Fusion 360 can be done by exporting a design to a 3D object file or through integration with Cura (for information on Cura, refer to *Chapter 2, What Are Slicer Programs?*).

TinkerCAD

TinkerCAD is another product from AutoDesk. TinkerCAD was founded in 2011 to make 3D modeling available for everyone. TinkerCAD is accessed through a web browser, with an extensive learning center available online as well.

In *Figure 4.6*, we can see a screenshot from a TinkerCAD instance. In this example, the **Text** component is used to create extruded text:

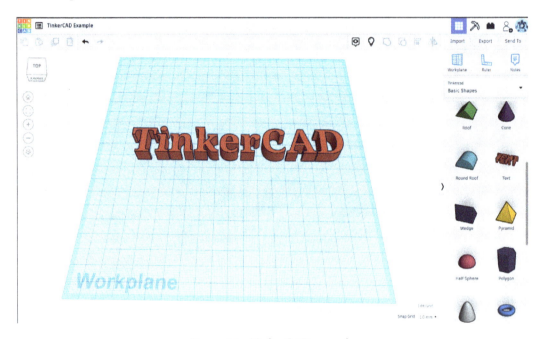

Figure 4.6 – TinkerCAD example

External objects may be imported onto the canvas of the program. Also, our design may be 3D printed directly from the app using several cloud-based platforms. For local printing, designs may be exported and downloaded as a 3D object.

For laser cutting, a bottom layer may be exported and stored as a `.svg` file. This can be useful in situations where a design requires a thick flat bottom plate to be cut by a laser or CNC router.

> **What Is a CNC Router?**
>
> CNC routers and 3D printers are similar in design as they share a flat plate area and a head that moves in the x and z directions. While 3D printers are often called additive manufacturing machines due to the way they add material to the build plate, CNC routers are referred to as subtractive manufacturing machines. The head on a CNC router cuts a block of material with a spinning router to create shapes.

TinkerCAD designs may also be exported for Minecraft and Lego brick templates.

FreeCAD

Like Fusion 360, **FreeCAD** offers an environment to create objects from sketches. As an open source program, FreeCAD has many external extensions from various developers. This gives FreeCAD a bit of a disjointed feel.

The workspaces available for FreeCAD are plentiful. In *Figure 4.7*, we can see one such workspace called the **Rocket** workspace. It is added through the add-on manager in FreeCAD:

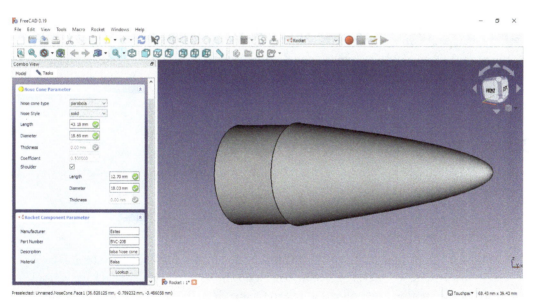

Figure 4.7 – Rocket nose cone design in FreeCAD

With the **Rocket** workspace, the user can create model rocket parts for 3D printing. A lookup table with access to model rocket parts from various manufacturers is available in the workspace. For the nose cone in *Figure 4.7*, the nose cone generated is the BNC-20B from Estes.

The latest version, as of the time of writing this book, is **FreeCAD 0.19.3**, indicating that FreeCAD is still very much in a beta release status. This should not dissuade any of us from trying it though as FreeCAD is feature-rich, and rivals many paid graphical CAD platforms. FreeCAD is free to download and use, and is available for Windows, macOS, and Linux.

Comparing OpenSCAD with other CAD programs

The most obvious difference between OpenSCAD and most CAD programs is the program-like interaction with the design. At first, this may seem like a disadvantage as it takes time to learn the commands; however, within a short time, this becomes an advantage.

Let's clarify. For example, to create a 10 mm x 10 mm x 10 mm cube in Fusion 360, we would create a sketch, draw a 10 mm x 10 mm box, and then extrude the box in the *z* direction by 10 mm. To do the same in OpenSCAD, we simply type `cube(10);` in the editor and hit *F5* to see the result.

Now that we are convinced of the power of OpenSCAD, let's take a closer look at it.

Learning OpenSCAD GUI and basic commands

It is time to get some hands-on experience. In this section, we will download and install OpenSCAD, look at the interface, and then create and view a few simple objects.

Let's start by installing OpenSCAD.

Downloading and Installing OpenSCAD

OpenSCAD is available for Windows, macOS, and Linux. To download and install OpenSCAD, perform the following steps:

1. In a browser, go to `http://openscad.org/downloads.html` to view the downloads available for OpenSCAD.
2. Follow the steps to install OpenSCAD for the appropriate operating system. For Windows users, we have a choice to download the installer or a zip package. Choose the installer option as it will put a link to OpenSCAD in our Windows Start menu.

Now that we have OpenSCAD installed, let's look at it in detail.

Getting to know the OpenSCAD environment

Let's now look at the OpenSCAD environment. To do so, open OpenSCAD and click on the **New** button. In the following screenshot, we can see what OpenSCAD looks like when it is opened with a blank design:

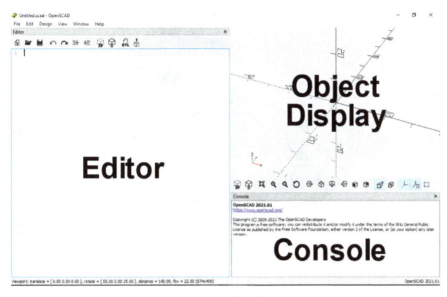

Figure 4.8 – OpenSCAD GUI environment

If the layout is not as shown in the preceding screenshot, then ensure that **Hide Error Log** and **Hide Customizer** are checked under the **Window** menu, and that **Hide Editor** and **Hide Console** are not checked under the **Window** menu.

As we can see in *Figure 4.8*, the screen is divided into three areas – **Editor**, **Object Display**, and **Console**. The only area that we may not hide is **Object Display** as this is the area that shows our designs. Before we take a more in-depth look at the OpenSCAD GUI, let's create our first design.

Creating our first design

Creating a simple design is extremely easy and quick in OpenSCAD. To create our first design, perform the following steps:

1. In **Editor**, type the following:

    ```
    sphere(10);
    ```

2. Hit *F5* on the computer keyboard or click on the **Preview** (as shown in *Figure 4.11*) button on the top menu.

3. Observe that we see a sphere in **Object Display**:

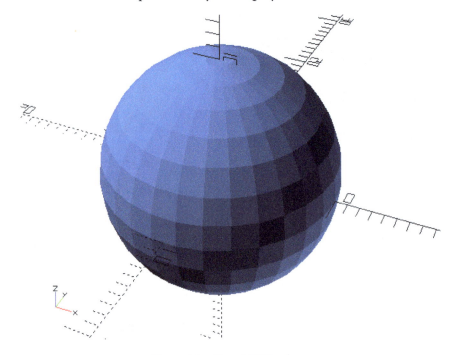

Figure 4.9 – OpenSCAD sphere

4. We have just created our first OpenSCAD design. Note how quick and easy it was. Also note how our sphere looks a little like a *disco ball*. To fix this, type in the following before the sphere command in the editor:

```
$fn=200;
```

5. Hit the *F5* key on the keyboard or click on the **Preview** button at the top. Observe that our object is now smooth:

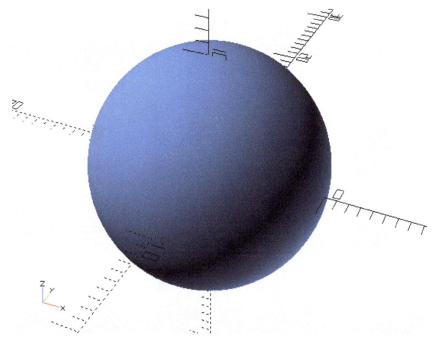

Figure 4.10 – Smooth OpenSCAD sphere

What Is a Disco Ball?

Disco balls, also known as mirror balls, are spheres with flat mirror pieces glued to them. The disco ball is usually attached to a motor that spins it around. Spotlights are projected onto the ball, creating a series of moving light circles around the room. Although some believe disco balls are a product of the 1970s (the height of disco music), their roots can be traced back to the 1920s.

The $fn keyword controls the number of fragments that are used to display our objects. Setting this to a high number (over 50) results in many fragments used at the cost of CPU memory. It is a good idea to keep this value low and increase it after the design is finalized.

Now that we have a little hands-on experience, let's take a closer look at the user interface.

Editor

In **Editor**, we can see 11 icons across the top (*Figure 4.11*). From the left to the right, we have the standard **New File**, **Open File**, **Save**, **Undo**, **Redo**, **Unindent**, and **Indent** buttons (make sure **Hide Editor toolbar** is unchecked under the **View** menu, if you cannot see the icons).

However, it is the four icons on the right that are of interest to us, as shown here:

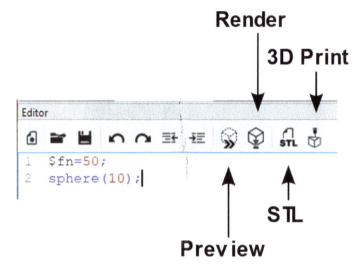

Figure 4.11 – OpenSCAD Editor

Let's take a closer look at these four buttons, starting with the **Preview** button:

- **Preview** – We use the **Preview** button, or *F5* on the keyboard, to generate a "preview" of our design.

- **Render** – We use the **Render** button, or *F6* on the keyboard, to render our design for export. Rendering takes longer than preview and uses a lot more computing resources.

- **STL** – We use the **STL** button, or *F7* on the keyboard, to export our design as a .STL file. Please note that this will only work for designs that have been rendered.

- **3D Print** – We use the **3D Print** button, or *F8* on the keyboard, to send our print either to the **Print a Thing** online service or OctoPrint (local network service for our 3D printer) on our network.

Now that we have a better understanding of the **Editor** pane, let's look at the area where we view our designs.

Object Display

We view our objects in the **Object Display** area. Across the bottom of this area, we see a row of buttons. The first two are a copy of the **Preview** and **Render** buttons from the **Editor** pane. The next ten buttons control how we view our object at various angles and magnifications:

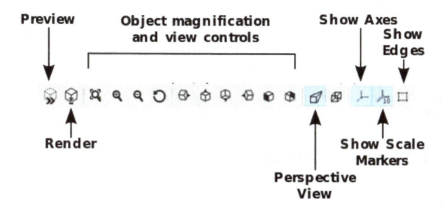

Figure 4.12 – Object Display buttons

As we can see in the *Figure 4.12*, there are three highlighted buttons indicating they are active. By default, our view is in **Perspective** mode with the **Show Axes** and **Show Scale Markers** buttons set as well. The final button allows us to view the edges of our object. In the case of our sphere, pressing this button will have no effect as the sphere does not have edges.

Along with the button functionality, we may use our mouse to move around in the **Object Display** area:

- *Left mouse button* – Holding down the left mouse button and moving the mouse allows us to rotate.

- *Right mouse button* – Holding down the right mouse button and moving the mouse allows us to pan.

- *Middle mouse button* – Holding down the middle mouse button (scroll wheel) and moving the mouse allows us to zoom in and out.

- *Scroll wheel* – Using the scroll wheel, we may zoom in and out without having to move the mouse.

Let's now look at **Console**.

Console

The **Console** area displays messages from OpenSCAD, such as the path of the design file on our computer. Compile errors are shown in the **Console** area as well. **Console** is the first place to look when there is an issue with a design.

Now that we have taken a look at the GUI, let's look into some of the basic shapes that we may create inside OpenSCAD.

OpenSCAD basic 2D shapes

Creating 2D shapes in OpenSCAD is relatively easy. Basic shapes, such as squares and circles, provide an excellent way to model real-world measurements. Using the `linear_extrude()` operation, we may turn 2D shapes into 3D objects.

Some basic 2D shapes include circles, squares, polygons, and text. We will use a couple of 2D shapes in the upcoming section, *Creating our PVC pipe hook*.

OpenSCAD basic 3D shapes

Basic 3D shapes in OpenSCAD include spheres, cubes, and cylinders. By combining basic 3D shapes, we can create complex objects. This, combined with extruded 2D shapes, is at the heart of an OpenSCAD design.

We will combine 3D objects and extrude 2D shapes in the upcoming section, *Creating our PVC pipe hook*.

Learning OpenSCAD Boolean and transformation operations

In this section, we will cover some of the basic Boolean and transformation operations. We will then use this knowledge to build a PVC pipe hook, as shown in *Figure 4.3*.

Let's start with Boolean operations.

OpenSCAD Boolean operations

There are three Boolean operations in OpenSCAD. These are as follows:

- `union()` – This operation joins shapes together. This can be used for 2D or 3D shapes (but not at the same time).
- `difference()` – This operation subtracts the second and subsequent shapes from the first shape. This operation may also be used on 2D and 3D shapes, but not at the same time.
- `intersection()` – This operation creates an intersection between shapes. Only the area in which the shapes overlap is retained. This may be used with 2D and 3D shapes as well, but not at the same time.

To demonstrate how each Boolean operation works, let's use a simple sketch. To do so, perform the following steps:

1. Open a new file in OpenSCAD by clicking on **File | New File**.

2. We will join a square and a circle together using the `union()` operation. Type the following code in the **Editor** pane:

```
$fn=50;
union()
{
    square(10, center = true);
    circle(d=11);
}
```

3. Click on the **Render** button, or hit the *F6* key. Observe the following shape:

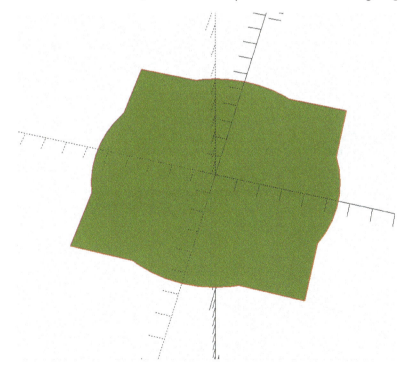

Figure 4.13 – Union of a square and a circle

4. Now that we see what union does, let's look at what difference does. In the code, change union() to difference(). Click on the **Render** button, or hit the *F6* key. Observe the following shape:

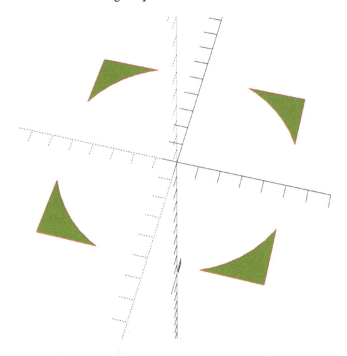

Figure 4.14 – Difference between a square and a circle

5. As we can see, the difference() operation subtracted the circle from the square. Now let's see what intersection does. In the code, change difference() to intersection(). Click on the **Render** button, or hit the *F6* key. Observe the following shape:

Figure 4.15 – Intersection between a square and a circle

As we can see, the intersection operation created a box-like shape with rounded edges. This shape represents the intersection or the area where both shapes overlap.

Now that we have explored Boolean operations, let's look at a few transformation operations.

OpenSCAD transformation operations

OpenSCAD transformation operations allow us to move, rotate, scale, resize, and mirror our objects. The following is a list of a few OpenSCAD transformation operations:

- `translate([x, y, z])` – This operation allows us to move an object in the **Object Display** area. Parameters passed are enclosed with square brackets.

- `rotate([x, y, z])` – This operation allows us to rotate an object in the **Object Display** area. The parameters x, y, and z represent the amount of rotation in their respective axes.

- `scale([x, y, z])` – This operation allows us to scale an object in the x, y, and z directions.

- `resize([x, y, z])` – This operation allows us to resize an object in the x, y, and z directions.

> **Don't Forget the Square Brackets**
>
> One mistake that is easy to make is the omission of the square brackets for the x, y, and z parameters. The best way to think about it is to consider the x, y, and z parameters as one value, and thus the need to group them together.

Now that we understand some of the commands in OpenSCAD, let's use it to create an object. We will create a PVC pipe hook, as shown in *Figure 4.3*.

Creating our PVC pipe hook

Using what we have learned to this point, it is time to start creating our own designs. The PVC pipe hook is a simple, yet practical, object. In *Figure 4.16*, we can see a version of the PVC pipe hook used to hold up a transmitter for a drone:

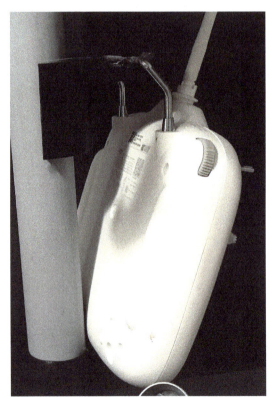

Figure 4.16 – PVC hook used for a drone transmitter

Let's start by creating our first shape.

Creating the first shape

To start off, we will use a cylinder as a base shape and then cut out a section for the PVC pipe. To do this, perform the following steps:

1. In the **Editor** pane, type in the following code:

    ```
    $fn=100;
    cylinder(d1=90, d2=100, h=60);
    ```

    The first line sets the quality of the image we see in the **Object Display** area, while the second line creates a 60 mm tall cylinder with a bottom diameter of 90 mm and a top diameter of 100 mm.

2. Click on the **Render** button, or hit the *F6* key, to see the cylinder.

3. We will now cut out a hole in the center of the cylinder equal to the diameter of the PVC pipe. For our example, we will cut for a PVC pipe with a diameter of 42.5 mm. To do so, change the code to the following:

    ```
    $fn=100;
    difference()
    {
            cylinder(d1=90, d2=100, h=60);

            translate([0, 0, -10])
            cylinder(d=42.5, h=100);
    }
    ```

4. Using the difference() operation, a cylinder with the same diameter as the PVC pipe is subtracted from the middle of the first cylinder. The height set for the PVC pipe does not matter for subtraction. The translate() operation is used on the PVC pipe cylinder to move it down so that it has a negative *z* axis value. This provides a clean cut at the bottom of our first cylinder. The value -10 is arbitrary. Click on the **Render** button, or hit *F6* on the keyboard, to see our new shape.

5. Observe that our cylinder now has a hole in it. We now need to cut it in half. To do this, we use a large cube. The size of the cube is irrelevant as long as it is larger than our cylinder. To do this, we change our code to the following:

    ```
    $fn=100;
    difference()
    ```

```
{
    difference()    {
        cylinder(d1=90, d2=100, h=60);


        translate([0, 0, -10])
        cylinder(d=42.5, h=100);
    }
    translate([-100,0,0])
    cube([200, 200, 200], center=true);
}
```

6. What we have done here is nest our original `difference()` operation inside another `difference()` operation. A large cube, centered and moved `-100` in the *x* axis (to put it on one side of the *x* axis) is subtracted from the first shape. Click on the **Render** button, or hit *F6* on the keyboard, to see our new shape.

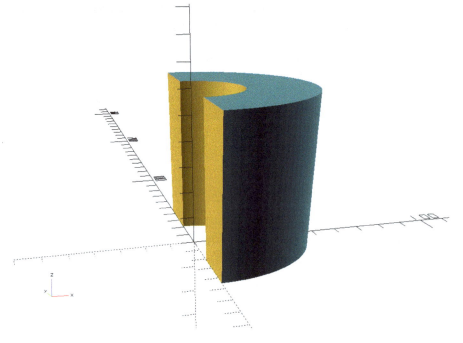

Figure 4.17 – PVC hook first shape

We can see our PVC pipe hook starting to resemble the hooks in *Figure 4.3*.

> **Using the Show Only (!) Symbol in OpenSCAD**
>
> As our designs become more complicated, it is easy to forget what a particular operation does. A technique we may use to help us isolate an operation is the Show Only command. The show only command is represented by the symbol ! and can be used to isolate an operation. For example, putting it in front of our cube will display the cube used to cut the cylinder in half: `!translate([-100,0,0])cube([200, 200, 200], center=true);`.

Now that we have the main body done, let's add some screw holes to the side.

Adding screw holes

To add screw holes, we subtract rotated cylinders. As our code is starting to get a little more complicated, we will comment out what we have done so far before we create the cylinders for the screw holes.

To do this, perform the following steps:

1. Highlight all the code except for the first line.
2. Click on **Edit | Comment**, or press *Ctrl + D* on the keyboard. Observe that the code is commented out:

```
$fn=100;
//difference()
//{
//     difference()     {
//          cylinder(d1=90, d2=100, h=60);

//          translate([0, 0, -10])
//          cylinder(d=42.5, h=100);
//     }
//     translate([-100,0,0])
//     cube([200, 200, 200], center=true);
//}
```

3.   We will now create two cylinders to use as cutaways for our screw holes. For our model, we will put these screw holes at 15 and 35 mm from the bottom. In the **Editor** pane, type the following:

```
translate([0,0,15])
rotate([0,90,0])
cylinder(d=3, h=100);


translate([0,0,35])
rotate([0,90,0])
cylinder(d=3, h=100);
```

4.   To understand OpenSCAD code, we read the statements from bottom to top. For the first statement, a 3 mm cylinder with a height of 100 mm is created, rotated 90 degrees along the *y* axis, and lifted 15 mm on the *z* axis. The second statement does the same, but lifts the cylinder 35 mm on the *z* axis. Click on the **Render** button, or hit *F6* on the keyboard, to see the cylinders.

5.   To add a professional look to our design, let's countersink our screws. To do this, we will add additional cylinders. Add the following lines of code:

```
translate([45,0,15])
rotate([0,90,0])
cylinder(d=10, h=50);


translate([45,0,35])
rotate([0,90,0])
cylinder(d=10, h=50);
```

6.   These two 10 mm diameter and 50 mm tall cylinders are rotated and placed at the same heights as the first two cylinders. The only difference is the *x* axis placement, which is set to 45 mm. This distance will create a countersink. As noted before, the height of the cylinders is irrelevant when they are used in subtraction. Click on the **Render** button, or hit *F6* on the keyboard, to see what we have so far.

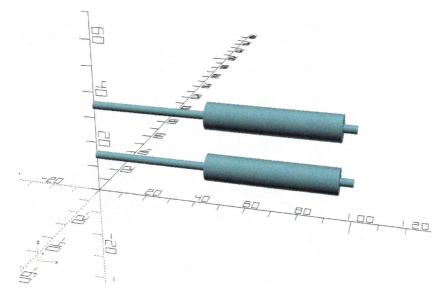

Figure 4.18 – Cutaways for countersink screw holes

7. For good measure, we should wrap our new lines of code in a union() operation. This will make it easier to keep track of. To do this, we change our code to the following:

```
union()
    {
        translate([0,0,15])
        rotate([0,90,0])
        cylinder(d=3, h=100);

        translate([0,0,35])
        rotate([0,90,0])
        cylinder(d=3, h=100);

        translate([45,0,15])
        rotate([0,90,0])
        cylinder(d=10, h=50);

        translate([45,0,35])
        rotate([0,90,0])
        cylinder(d=10, h=50);
    }
```

8. Now that we have our screw hole cylinders created, it is time to use them to create the screw holes. To do this, we use another difference() operation to subtract the screw hole cylinders from the first shape we created. Uncomment out the previous code and wrap it with our newer code like this:

```
$fn=100;
difference()
{
    difference()
    {
        difference()    {
            cylinder(d1=90, d2=100, h=60);

            translate([0, 0, -10])
            cylinder(d=42.5, h=100);
        }

        translate([-100,0,0])
        cube([200, 200, 200], center=true);
    }
    union() {
        translate([0,0,15])
        rotate([0,90,0])
        cylinder(d=3, h=100);

        translate([0,0,35])
        rotate([0,90,0])
        cylinder(d=3, h=100);

        translate([45,0,15])
        rotate([0,90,0])
        cylinder(d=10, h=50);

        translate([45,0,35])
        rotate([0,90,0])
        cylinder(d=10, h=50);
    }
}
```

9. Essentially, what we are doing here is subtracting the screw hole cutaways from our first object. Click on the **Render** button, or hit *F6* on the keyboard, to see what we have so far:

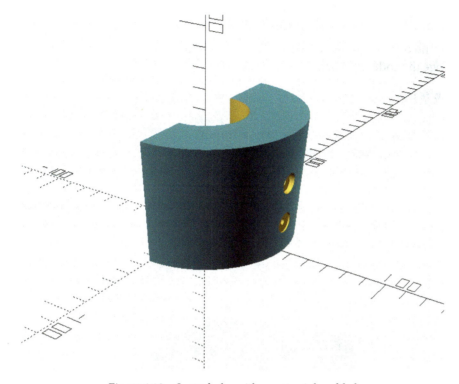

Figure 4.19 – Screw holes with countersinks added

We are now ready to add the hook part. We will create 2D shapes in OpenSCAD to accomplish this.

Creating the hook

To create the hook, we will start by combining a box and a circle. We will then extrude the shape and move it to the top of the cylinder. To do this, perform the following steps:

1. Comment out all of the code so far by selecting it and clicking on **Edit | Comment**, or type *Ctrl* plus *D* on the keyboard.

2. In the **Editor** pane, type in the following:

```
translate([30, 0])
square([60, 70], center=true);
```

3. Getting the size of the square right is a trial-and-error process. With OpenSCAD, we can easily change values while we work on our design. We are not committed to the first shape created. In our code, a square with an *x* value of 60 and a *y* value of 70 will work well for our hook. Click on the **Render** button, or hit *F6* on the keyboard, to see our initial square.

4. To round off the square, we will add a circle and use the intersection operation. Change the code from *Step 2* to the following:

```
intersection()
{
    translate([20,0])circle(d=80);
    translate([30,0])square([60, 70], center=true);
}
```

5. A circle with a diameter of 80 mm, moved 20 mm in the *x* direction, provides a good intersection for the square. Click on the **Render** button, or hit *F6* on the keyboard, to see what we have so far:

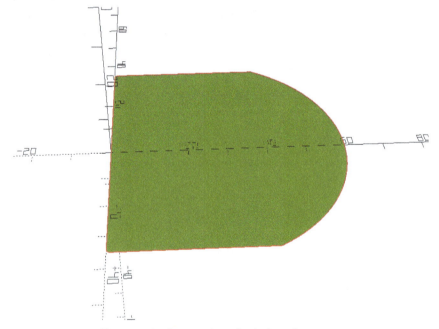

Figure 4.20 – Intersection of a circle and a square

6. As we can see, the shape is a little more interesting than a simple square. We may change the values at any time during our design to fine-tune it. With our desired shape, it is time to change our 2D design to a 3D design, and move the shape into place. We will use a `linear_extrude()` operation and a `translate()` operation to do this. Change the code from *Step 4* to the following:

```
translate([25,0,45])
linear_extrude(15)
intersection()
{
    translate([20,0])circle(d=80);
    translate([30,0])square([60, 70], center=true);
}
```

7. The `linear_extrude()` operation turns our 2D shape into a 15 mm thick 3D shape, and the `translate()` operation moves it into the proper place for our design. Uncomment out the previous code. Click on the **Render** button, or hit *F6* on the keyboard, to see the design up to this point:

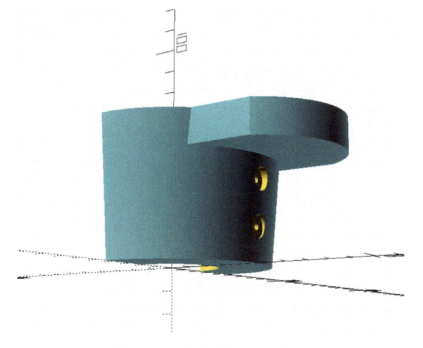

Figure 4.21 – Hook added to the design

8. The only thing left to do is to cut away a half-circle at the top of the hook. We do this by subtracting a cylinder put in the proper place. For our final design, the first part of our code builds the base cylinder shape:

```
$fn=100;
difference()
{
  union()
  {
    difference()
    {
      difference()
      {
        difference()
        {
          cylinder(d1=90, d2=100, h=60);

          translate([0, 0, -10])
          cylinder(d=42.5, h=100);
        }
        translate([-100,0,0])
        cube([200, 200, 200], center=true);
      }
      union()
      {
        translate([0,0,15])
        rotate([0,90,0])
        cylinder(d=3, h=100);

        translate([0,0,35])
        rotate([0,90,0])
        cylinder(d=3, h=100);

        translate([45,0,15])
        rotate([0,90,0])
        cylinder(d=10, h=50);
```

```
        translate([45,0,35])
        rotate([0,90,0])
        cylinder(d=10, h=50);
    }
}
```

9. The second part of our code adds the hook and creates the groove at the top of the hook:

```
translate([25,0,45])
    linear_extrude(15)
    intersection()
    {
        translate([20,0])circle(d=80);
        translate([30,0])square([60, 70], center=true);
    }
}
    translate([60,50,70])
    rotate([90,0,0])
    cylinder(d=30, h=100);
}
```

10. Click on the **Render** button, or hit the *F6* key, to see our completed design:

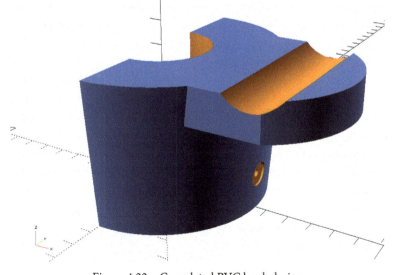

Figure 4.22 – Completed PVC hook design

This completes our design. As we can see, the code for even simple objects can become quite complex. In the coming chapters, we will organize our code into modules and libraries to make it easier to understand and maintain.

Summary

In this chapter, we introduced the CAD design software, OpenSCAD. We discussed some of its features and compared it to other CAD environments on the market. We then became familiar with its graphical interface by looking at the major components.

We started writing code to generate our first design and learned a little bit about the types of objects available in OpenSCAD. We took a hands-on approach to learning by creating a simple sphere. We looked at the Boolean and transformation operations and demonstrated these concepts with examples. We finished off the chapter by creating a PVC pipe hook.

In the next chapter, we will take what we have learned so far and dive a little deeper into more complex OpenSCAD coding.

5
Using Advanced Operations of OpenSCAD

Based on what we initially saw while loading OpenSCAD, we may have underestimated its power. However, by using advanced operations, we can create dynamic design code – or, in other words, code that can easily be reused for multiple designs.

OpenSCAD lets you add and extrude text for use in creating designs that can be used in mass customization business models – an absolute strength of 3D printing. With mass customization, we can tailor our product specifically to our client at a scale that was not possible in the days before 3D printers.

In this chapter, we will cover the following topics:

- Turning 2D shapes into 3D objects
- Looking at advanced OpenSCAD commands
- Simplifying our code with modules

Technical requirements

The following will be required to complete the chapter:

- Any late model Windows, macOS, or Linux computer that can install OpenSCAD

- The code and images for this chapter, which can be found at `https://github.com/PacktPublishing/Simplifying-3D-Printing-with-OpenSCAD/tree/main/Chapter5`.

Turning 2D shapes into 3D objects

Part of constructing our PVC Hook in *Chapter 4*, *Getting Started with OpenSCAD*, involved using the `linear_extrude` command to turn a 2D shape into a 3D shape. We created the 2D shape in OpenSCAD using the shapes it had available.

Although the result was what we needed, there will be times when we may require a shape that is a little difficult to create in OpenSCAD. For these situations, we must import a `.svg` file and then extrude it.

In this section, we will do just that. We will create a 3D printable Thumbs Up award, import it as a `.svg` file, and then extrude it.

Importing SVG files into OpenSCAD

To import a `.svg` file into OpenSCAD, we can use the `import` command. To import the Thumbs Up graphic for our design, do the following:

1. Download the `ThumbsUp.svg` file from the following GitHub location: `https://github.com/PacktPublishing/Simplifying-3D-Printing-with-OpenSCAD/tree/main/Chapter5/images`.

2. Create a new OpenSCAD file called `thumbs-up-award.scad` and save it in the same location as the `ThumbsUp.svg` file.

3. Type in the following line in the Editor:

```
import("ThumbsUp.svg", center=true);
```

4. Save the file. Observe the following shape in the Object Display:

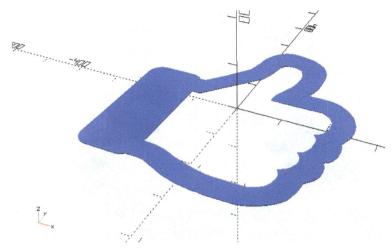

Figure 5.1 – Importing ThumbsUp.svg

5. If the shape cannot be seen, scroll out or click on **View All** (*Ctrl + Shift + V*).

Although seeing a large thumbs-up sign in our Object Display is impressive, there is not much we can do with it, in its current form. We will have to extrude it to make it a 3D shape, and then add a base to turn it into an award that can sit on a desk.

Let's do this now.

Creating a 3D Thumbs Up symbol

Using the center=true option makes working with a graphic file much easier, as we know we will have the same starting position for any additional shapes we add. Before we can add more shapes, we must turn our graphic into a 3D shape. To do so, follow these steps:

1. Add a linear_extrude command before the imported graphic. To keep the code clean, we will use separate lines:

```
linear_extrude(10)
import("ThumbsUp.svg", center=true);
```

By passing in the value of 10 to the linear_extrude command, we are telling the OpenSCAD compiler to extrude our shape out by 10 mm.

2. Click on the **Render** button or hit *F6* on your keyboard. Observe that our Thumbs Up graphic is now a 3D shape:

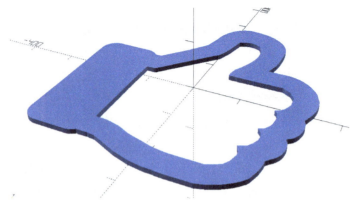

Figure 5.2 – Thumbs Up after using the extrude operation

3. To make it easier to work with our new shape, let's rotate it so that we can add a base under it. Add the following `rotate` command before the `linear_extrude` command:

```
rotate([90,0,0])
```

4. Click on the **Render** button or hit *F6* on your keyboard. Observe that an extruded Thumbs Up object appears in the Object Display and is rotated along the *x* axis:

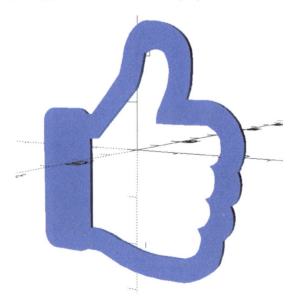

Figure 5.3 – Extruded Thumbs Up graphic

Now that we have an extruded shape, let's add a base to it.

Extruding the base

Adding a base is relatively simple. We could use a 3D cylinder, but for our design, we will create our base as a 2D shape, and then use the `rotate_extrude` command to turn it into a 3D shape. To do this, follow these steps:

1. We will start by commenting out the existing code. Highlight the code that's been written and hit *Ctrl + D* on your keyboard.

2. For our initial shape, we will subtract a circle from a square using the `difference` operation. In the Editor, type the following:

```
difference()
{
    translate([100,0])square(200, true);
    translate([200,0])circle(80);
}
```

3. Click on the **Render** button or hit *F6* on your keyboard. Observe that our initial shape looks as follows:

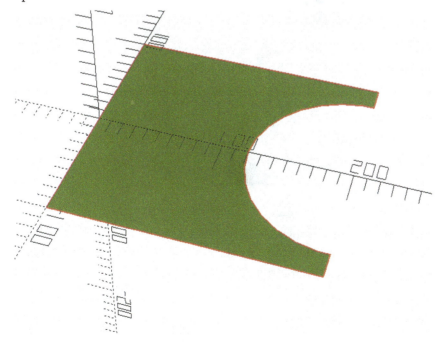

Figure 5.4 – Initial base shape

4. To turn our 2D shape into a 3D shape, we will use the `rotate_extrude` operation. Using this operation, our 2D shape will be rotated along the *y* axis, and then the new shape will rotate 90 degrees along the *x* axis. Add the `rotate_extrude` operation before the `difference` operation so that our code looks like this:

```
rotate_extrude(angle=360)
difference()
{
    translate([100,0])square(200, true);
    translate([200,0])circle(80);
}
```

The `rotate_extrude` operation takes a parameter called `angle`. This specifies the amount of rotation that is applied to our 2D shape. In our case, we would like it to be a closed shape, so we passed in a value of `360` degrees.

5. Hit *F6* or click the **Render** button. Observe that our 2D shape is now a 3D shape and has been rotated 90 degrees in the *x* direction:

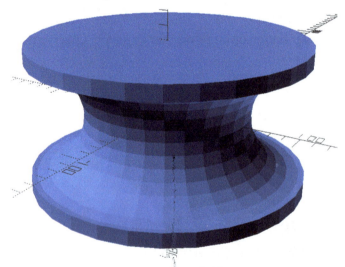

Figure 5.5 – Extruded base

6. Although our shape does look like a base, it would be nice to stretch it vertically to give it a taller look. We could change the parameters of our 2D rectangle and circle; however, an easier solution is to stretch our shape with the `scale` operation. Add the `scale` operation before the `rotate_extrude` operation so that our code looks like this:

```
scale([1,1,3])
rotate_extrude(angle=360)
```

```
difference()
{
    translate([100,0])square(200, true);
    translate([200,0])circle(80);
}
```

The `scale` operation takes *x*, *y*, and *z* scale factors as one parameter (enclosed in square brackets). By setting the value to [1,1,3], we are keeping the same size in the *x* and *y* directions but stretching it three times in the *z* direction.

7. Hit *F6* or click the **Render** button. Observe that our base is now taller:

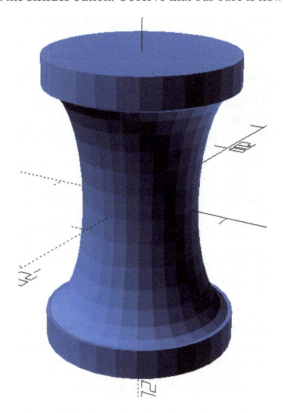

Figure 5.6 – Base after using the scale operation

Now that we have created the base, it is time to uncomment out the Thumbs Up object and place it on the base.

Putting the base and Thumbs Up object together

We will now uncomment out the Thumbs Up object and move it up in the *z* direction so that it will fit neatly on top of the base. To do this, follow these steps:

1. Select and uncomment out the code that was commented out in *Step 1* of the *Extruding the base* section.

2. To make our base smoother, put the following code at the top:

    ```
    $fn=200;
    ```

3. Now, we will move our Thumbs Up object up in the *z* direction so that it will fit on the base. To do this, we will use the `translate` operation just before the `rotate` operation. Our code should look like this:

    ```
    $fn=200;
    translate([0,0,545])
    rotate([90,0,0])
    linear_extrude(10)
    import("ThumbsUp.svg", center=true);

    scale([1,1,3])
    rotate_extrude(angle=360)
    difference()
    {
        translate([100,0])square(200, true);
        translate([200,0])circle(80);
    }
    ```

    Using the `translate` operation, we can move the Thumbs Up award up by the value of `545` (this value is determined by trial and error). This places the Thumbs Up object neatly on top of the base.

4. Hit *F6* or click the **Render** button. Observe that our object looks as follows:

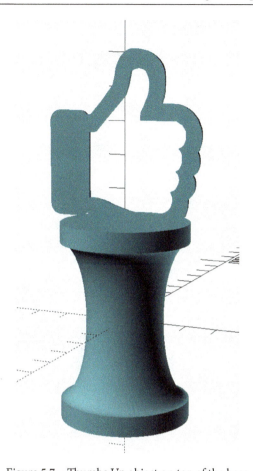

Figure 5.7 – Thumbs Up object on top of the base

As we can see, the Thumbs Up object sits just on top of the base. Our completed design can now be 3D printed.

However, if we want to 3D print our award, we will require support material for the empty area inside the hand. This would have to be removed and may damage our print. A better way to print our Thumbs Up award would be to divide it into two pieces and glue them together. Let's do just that.

3D printing our Thumbs Up award

The logical division for our award would be where the base and the Thumbs Up object intersect. Cutting a groove that fits the Thumbs Up object into the base would make gluing easier.

To create a base with a cut-out, we can use the `difference` operation on the base:

1. To create the cut-out, change our code to the following:

```
$fn=200;
difference()
{
    scale([1,1,3])
    rotate_extrude(angle=360)
    difference()
    {
        translate([100,0])square(200, true);
        translate([200,0])circle(80);
    }
    translate([0,0,545])
    rotate([90,0,0])
    linear_extrude(10)
    import("ThumbsUp.svg", center=true);
}
```

> **Using Modules for our Code**
>
> As code starts to grow, it is a good idea to wrap it inside modules. Modules allow us to simplify our code into one line since calling a module replaces many lines. We will be implementing modules in the *Simplifying our code with modules* section.

We modified the code to put the base first as we want to cut out the Thumbs Up object from the base to create the groove.

2. Hit *F6* or click the **Render** button. Observe that the base now has a groove cut out:

Figure 5.8 – Base with groove

3. To create just the Thumbs Up object, we will return to using the `linear_extrude` operation on the imported `.svg` file. Open a new file in OpenSCAD and put in the following code:

```
linear_extrude(10)
import("ThumbsUp.svg", center=true);
```

4. Hit *F6* or click the **Render** button. Observe the extruded Thumbs Up object shown in *Figure 5.2*.

In *Chapter 7*, *Creating a 3D-Printed Name Badge*, we will start printing out our designs. For now, it is enough to know that we should always consider how an object will be 3D printed before we start our design.

By breaking our Thumbs Up award into two pieces, we make it easier to print. The following screenshot shows how we would print out our object using Cura:

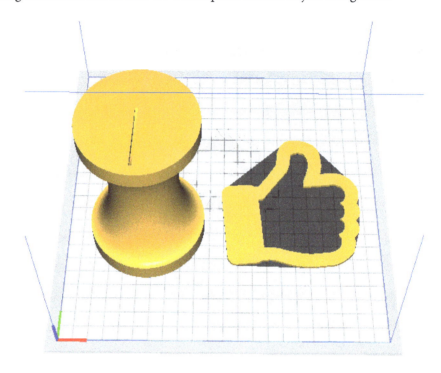

Figure 5.9 – Thumbs Up award in Cura

The objects shown in the preceding screenshot have been scaled to 20% to have them fit on the build plate of our Ender 3 V2. For reference, the base will print out with a diameter of 80 mm and a height of 120 mm.

As we can see, it would be relatively easy to glue our Thumbs Up award together after printing. The groove in the base provides a friction fit for the Thumbs Up object to hold it while the glue dries.

Taking Layer Lines into Account

When we print out our designs, we should always take the layer lines into account. An object will always be weaker along layer lines due to the way **fused deposit modeling** (**FDM**) 3D printing works. By printing the Thumbs Up object flat, we not only reduce the printing time but also make the object much stronger.

Although most of us would have limited use for a Thumbs Up award, it is easy to see how the concepts we've used here to create the award may be applied elsewhere. Having the ability to import .svg files and turn them into 3D objects allows us to create detailed designs that would be difficult to create any other way.

Now that we have a grasp of turning 2D shapes into 3D objects in OpenSCAD, let's look at some of the more complex OpenSCAD operations.

Looking at advanced OpenSCAD commands

A quick and extensive reference for OpenSCAD commands is the OpenSCAD cheat sheet, which can be accessed from the **Help** menu (click on **Help | Cheat Sheet**). As we can see, there are quite a few operations and commands that we can add to our OpenSCAD scripts.

In this section, we will look at a few of these operations to create a plaque for our Thumbs Up award. Specifically, we will look at the text and len operations.

We will start by looking at the fonts that are available for the text operation. For our purposes, we will be using a monospaced font.

What Are Monospaced Fonts?

Those of us who are old enough to remember a world where correspondence was done using typewriters are already familiar with monospaced fonts. If we think of the way a typewriter punches letters onto paper, we can visualize that the space between each letter is the same (monospaced). With the introduction of word processors, proportional fonts became possible for creating text that is more visually appealing.

Let's take a look at the fonts we can use in our design.

Exploring the available fonts

With OpenSCAD, we do not need to install extra fonts to use the text operation. We can simply call upon the fonts that have already been installed in our operating system using the OpenSCAD **Font List** dialog. We can use this tool to copy the desired font name to our clipboard and paste it into our program.

To do so, follow these steps:

1. In OpenSCAD, click on **Help | Font List**. The **OpenSCAD Font List** dialog will open.

2.   Scroll down the list and select the **Courier New** font. If this font is not available, select another monospaced font, such as **Lucida Sans Typewriter**:

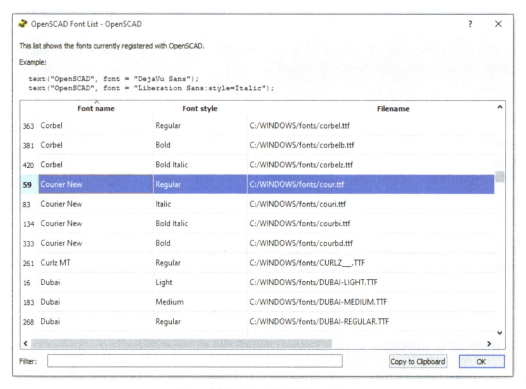

Figure 5.10 – OpenSCAD Font List

3.   To copy the font name to our clipboard, click on the **Copy to Clipboard** button.

4.   Click **OK** to close the dialog.

Now that we have copied the font name, it is time to use it in an OpenSCAD program. To do this, we will use the text operation to create a plaque for our Thumbs Up award.

Exploring the text operation

The text operation in OpenSCAD is easy to use. It takes in a string of text to be displayed, as well as other parameters such as size and the font name. To use the text operation, follow these steps:

1.   In a new OpenSCAD file, type in the following:

```
text("Good Job!", font=
```

2. Position the cursor to the right of the = sign and paste the contents of the clipboard. Close the statement with a right round bracket and semi-colon. The code should look as follows:

```
text("Good Job!", font="Courier New:style=Regular");
```

3. Hit *F6* or click the **Render** button. Observe that the text **Good Job!** is displayed:

Figure 5.11 – Output of the text operation

4. As OpenSCAD is a programming environment, we can use variables as substitutions for values. For our case, this would be the text we display on our plaque. Using a variable, named `display_text`, change the code to the following:

```
display_text="Good Job!";
text(display_text, font="Courier New:style=Regular");
```

What we have done here is substitute the **Good Job!** string in the `text` operation with a variable set to the same value.

5. Hit *F6* or click the **Render** button. Observe that we get the same result that's shown in the preceding screenshot.

6. To make use of our text, we need to convert it into 3D. To do that, we can use the `linear_extrude` operation. We will extrude our text by 5 mm. Modify the code to the following:

```
display_text="Good Job!";

linear_extrude(5)
text(display_text, font="Courier New:style=Regular");
```

7. Hit *F6* or click the **Render** button. Observe that our text is now 3D:

Figure 5.12 – Extruded text

As we can see, adding text to our OpenSCAD designs is not difficult. Substituting variables for display values is a good practice as variables are generally set at the beginning of a program and are easy to find. This allows the programmer to change display values quickly.

Now that we understand the `text` operation, we will create a plate that the text will sit on.

Creating a dynamic backing plate

To create a dynamic backing plate, we need to know the size of the text we are displaying. We purposely used a monospace font so that we know that to determine the size of the plate, we simply need to count the number of characters and multiply it by the size of one character. We will count the number of characters using the `len` operation and display the number with the `echo` operation.

To do this, we must modify the code from the previous section:

1. Modify the code from the *Exploring the text operation* section so that it looks like this:

```
display_text="Good Job!";


linear_extrude(5)
text(display_text, font="Courier New:style=Regular");


echo(len(display_text));
```

2. Using the `echo` operation, we send information to the console. In our case, we are interested in the length of the string that's stored in the `display_text` variable. We use `display_text` to store the `Good Job!` string, which is displayed in our design. The `len` operation returns the length of the string that's stored in the `display_text` variable. Hit *F6* or click the **Render** button. Observe the text `ECHO` in the console, followed by the number 9:

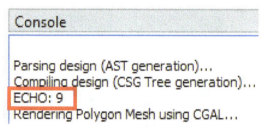

Figure 5.13 – Echo message in the console

3. The number 9 represents the number of characters in the `Good Job!` string, which is stored in the `display_text` variable. To verify that `len` and `echo` work properly, change the value of `display_text` to the following:

```
display_text="Very Good Job!";
```

4. Hit *F6* or click the **Render** button. Verify that the value following `ECHO` has changed to 14, which corresponds to the number of characters in the `Very Good Job!` string.

5. Now that we know how to retrieve the number of characters in a string, it is time to use this information to build a backing plate. To do so, let's build a plate around a single character first so that we know what size plate we need. Change the value of `display_text` to the following:

```
display_text="A";
```

6. Setting `display_text` to `A` will allow us to find the correct size for a single character. Since we are using a monospaced font, this value will be the same, regardless of which character we use. The `A` value is arbitrary. Add the following to the bottom of the code:

```
cube([8.3,9,2]);
```

7.   This code adds a cube to our design with an *x* value of 8.3, a *y* value of 9, and a *z* value of 2. Hit *F6* or click the **Render** button. Verify that our design looks as follows:

Figure 5.14 – Letter A extruded with a backing plate

8.   Looking closely at the backing plate, we can see that it sits right in the middle of the letter A. You can make changes to both the *x* and *y* values to make any adjustments. Now that we have the correct size for our character, it is time to use the `len` operation to create a dynamic backing plate. Change the code from *Step 6* to the following:

```
cube([len(display_text)*8.3,9,2]);
```

9.   This code will check the length of `display_text` and multiply it by the size of a single character (8.3). Change the value of `display_text` back to the original text. Our completed code should look as follows:

```
display_text="Good Job!";


linear_extrude(5)
text(display_text, font="Courier New:style=Regular");
echo(len(display_text));


cube([len(display_text)*8.3,9,2]);
```

10. Hit *F6* or click the **Render** button. Verify that the backing plate covers all the text:

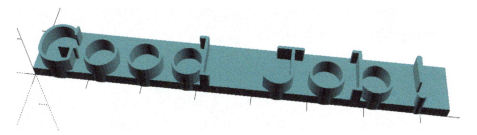

Figure 5.15 – Dynamic backing plate

11. To use our dynamic backing plate code with different text, we can simply change the value of display_text and render our design. Let's do just that. Change the value of display_text to the following:

```
display_text="Best in Class!";
```

12. Hit *F6* or click the **Render** button. Verify that our design now looks as follows:

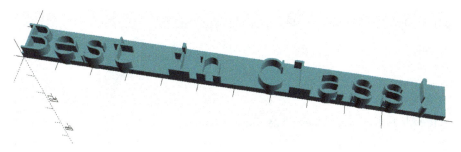

Figure 5.16 – Rendered Best in Class plaque

As we can see, writing our code with built-in dynamic elements allows us to quickly modify our designs. However, the more dynamic elements we add, the more complex our code becomes. This is where modules come in handy. In the next section, we will simplify our code by writing modules.

Simplifying our code with modules

As we've seen, OpenSCAD code can start to become quite complex. This not only makes it more difficult to maintain but makes our coding prone to errors. An elegant way to deal with this is to break our code into modules. Although we can put any code we want into a module, it is best practice to keep a module limited to a single function. For example, a good way to break up the Thumbs Up award would be by using code to create the base, code to create the Thumbs Up symbol, and code to create the plaque.

The syntax to create a module in OpenSCAD is the word `module`, followed by opening and closing parenthesis and open and closing curly braces:

```
module name_of_module(parameters)
{
    body_of_module
}
```

As we can see, modules are similar in their syntax to the `difference`, `union`, and `intersection` operations. It is a good idea to name the modules with verbs since they perform actions.

We will start exploring modules by enclosing the code to create the Thumbs Up object.

Creating a module for our Thumbs Up object

To encapsulate the code to create the Thumbs Up object in a module, we can simply wrap the code with the proper syntax. To do this, follow these steps:

1. In a new OpenSCAD file, create a module using the following code:

    ```
    module create_thumbs_up()
    {
    }
    ```

2. Save the file to the same location as the `ThumbsUp.svg` file.

3. Insert this code to create the Thumbs Up object between the curly braces so that it looks as follows:

    ```
    module thumbs_up()
    {
        rotate([90,0,0])
        linear_extrude(10)
        import("ThumbsUp.svg", center=true);
    }
    ```

4. If we were to try and render our design at this stage, we would not see anything as we have only defined the module and not called it. To implement the module, we must call it as if it were a built-in operation in OpenSCAD. Put the following code below the module code:

    ```
    thumbs_up();
    ```

5. Hit *F6* or click the **Render** button. Verify that we see the Thumbs Up object in the Object Display, as shown in *Figure 5.3*.

6. Although our module works as intended, it is very specific to the `ThumbsUp.svg` file. It would be nice to make our module a bit more dynamic and have it extrude any `.svg` file. To do so, we may take advantage of the ability to pass parameters into a module. Change the code to the following to make our module more dynamic:

```
module create_3D_from_svg(path)
{
    rotate([90,0,0])
    linear_extrude(10)
    import(path, center=true);
}
create_3D_from_svg("ThumbsUp.svg");
```

7. Hit *F6* or click the **Render** button. Verify that the Thumbs Up object is shown in the Object Display, as shown in *Figure 5.3*.

8. By giving our module a more generic name and passing in the path to our .svg file, we can use the module for whichever 2D file we choose. This makes our code more dynamic. To clean up the Editor, let's close the body of the module by clicking on the – symbol:

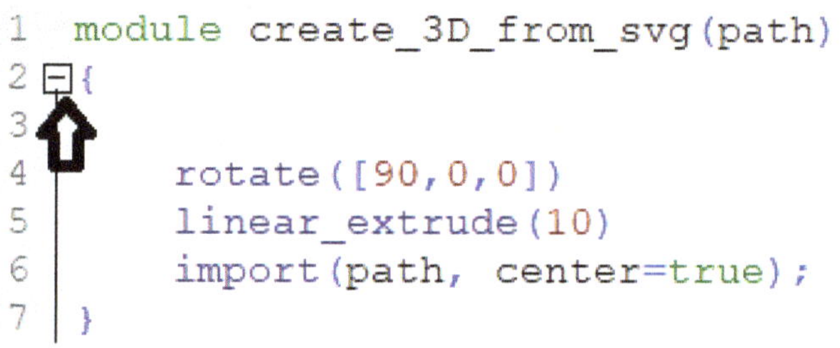

```
1  module create_3D_from_svg(path)
2 ⊟{
3
4        rotate([90,0,0])
5        linear_extrude(10)
6        import(path, center=true);
7  }
```

Figure 5.17 – The create_3D_from_svg module

As we can see, modules help in simplifying our code by breaking it into more manageable pieces. Now that we have the Thumbs Up object creation encapsulated in a module, let's turn our attention to the base.

Creating a module for the base

For our base, there are a few parameters that we can make dynamic. The most logical would be to parameterize the scale factors. However, we will leave the scale factors alone for now and focus on modifying the shape. We will use a default value to create a default shape.

To do this, follow these steps:

1. Comment out the following line so that we can start with a blank Object Display:

    ```
    //create_3D_from_svg("ThumbsUp.svg");
    ```

2. One of the benefits of writing our code with modules is not having to comment out a lot of code when we want to try something new. We simply need to comment out the call to the module. Place your cursor above the line we just commented out and type in the following:

    ```
    module create_base(shape=1)
    {
        scale([1,1,3])
        rotate_extrude(angle=360)
        difference()
        {
            translate([100,0])square(200, true);
            translate([200,0])circle(80 * (1/shape));
        }
    }
    create_base();
    ```

    By using a default parameter (shape=1), we can call our module without parameters. Here, we are using the shape variable to modify the radius of the circle, and thus the overall shape of our base.

3. Hit *F6* or click the **Render** button. Verify that our base looks the same as what's shown in *Figure 5.6*.

4. Now, let's change the shape of our base by passing in a number other than 1. Change the create_base() call to the following:

    ```
    create_base(2);
    ```

5.   Hit *F6* or click the **Render** button. Observe that the shape of the base has changed:

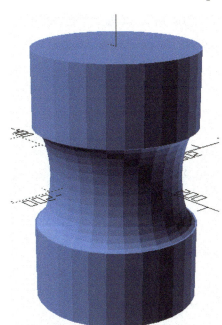

Figure 5.18 – Modified base

Now that we have a module for the base, it is time to turn our attention to the plaque. In the next section, we will create a module that will generate the plaque for our award.

Creating a module for the plaque

For our plaque, we know that the text that's displayed is dynamic. We could make the font dynamic as well; however, we would be essentially limited to monospaced fonts.

If we compare the size of the base to the plaque, we can see that the base is far larger. This is due to the size of the original .svg file that we imported as we matched the base to fit it. Thus, a parameter we should consider for our plaque module is a scale factor.

To make our plaque a bit more visually appealing, we should add more space above and below the letters. Also, centering the plaque on the *x* axis will make it easier to move around. We can achieve both these things using the translate operation.

To write our module, follow these steps:

1. Comment out the `create_base();` line.

2. Add the following code below the existing modules:

```
module create_plaque(display_text, scale_factor=1)
{
    size=len(display_text)*(8.3);

    scale([scale_factor,scale_factor,scale_factor])
    translate([-size/2,0,0])
    union()
    {
        linear_extrude(5)
        text(display_text,
         font="Courier New:style=Regular");

        translate([0,-9,0])
        cube([size,20,2]);
    }
}
```

What we are doing here is passing in the plaque text, as well as a scale factor. We are using a default value of 1 for the scale factor so that we do not need to set this value if we are not changing the size of our plaque. The `size` variable is used to store the size of our plaque. It is determined by taking the length of the text that's passed in and multiplying it by 8.3 (the size of a character). By knowing `size`, it is easy to determine how far to move our plaque negatively in the x direction to center it as it would be half of `size`. A `translate` operation (`translate([0,-9,0])`) is added to account for a larger y size of the cube (20) than what was provided in the *Creating a dynamic backing plate* section. This results in space above and below the text to give it a nicer look with a larger bottom section, as this will be where the plaque is inserted into the base.

3. Hit *F6* or click the **Render** button. Observe that the plaque has been created and is in the center of the Object Display:

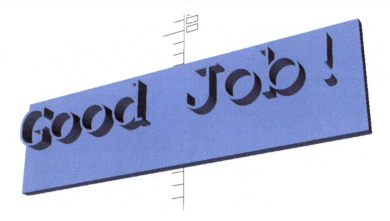

Figure 5.19 – Plaque rendered using a module

Now that we have the three modules, we will put them together and redesign our Thumbs Up award.

Creating a design using modules

To start our design, it is a good idea to minimize the modules that we've created and delete the code that calls them. Our Editor should look as follows:

```
Editor

 1  module  create_3D_from_svg(path)
 2 ⊞{
11
12  module  create_base(shape=1)
13 ⊞{
24  module  create_plaque(display_text,  scale_factor=1)
25 ⊞{
33 |
```

Figure 5.20 – Modules minimized

Now, we will start to code our design. We will use modules to create the shapes and the `translate` operation to move the shapes into place. To do this, follow these steps:

1. We will start by creating the base. Below the modules, type in the following in the Editor:

    ```
    $fn=200;
    create_base(2);
    ```

2. With this code, we are creating a base with a shape that's been modified by passing in the number 2 (as opposed to 1, so that we can create a thicker base than the one shown in *Figure 5.6*). Normally, we would put the `$fn=200;` line at the top of our program; however, it will serve well in this spot as we can easily change it if rendering is taking too long. Hit *F6* or click the **Render** button. You will see that the base has been created:

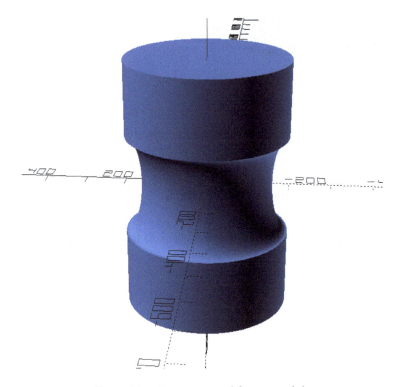

Figure 5.21 – Base generated from a module

3. To create the Thumbs Up symbol and place it in the right spot (as we did in the *Putting the base and Thumbs Up object together* section), we can use the `translate` operation on the `create_3D_from_svg()` module. Add the following code:

```
translate([0,0,545])
create_3D_from_svg("ThumbsUp.svg");
```

4. Hit *F6* or click the **Render** button. Observe that the Thumbs Up symbol appears on top of the base, as it did in *Figure 5.7*:

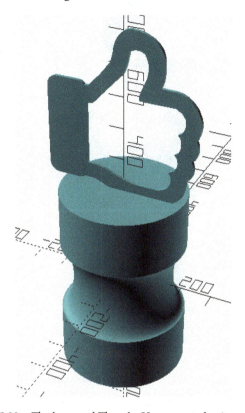

Figure 5.22 – The base and Thumbs Up generated using modules

5. Now that we have the base and Thumbs Up object in place, it is time to create the plaque and put it in place. Add the following code:

```
translate([0,-10,310])
rotate([90,0,0])
create_plaque("Good Job!",4);
```

What we are doing here is creating a plaque with the text `"Good Job!"` at a four times scale, rotating it, and then moving it into place. The values we use in the `translate` operation are derived from trial and error.

6. Hit *F6* or click the **Render** button. Observe the completed design:

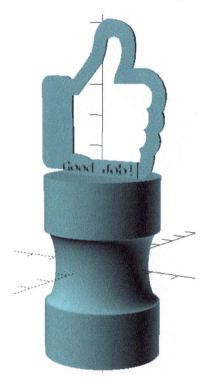

Figure 5.23 – The completed design

As we can see, dividing our code into modules makes it easier to create designs in OpenSCAD. Once a module has been defined, we can minimize it to create space in the Editor. Writing our modules to accept parameters makes them dynamic, which leads to more reusable code.

Summary

In this chapter, we dived deeper into the functionality of OpenSCAD. We learned how to create a 3D design from a 2D `.svg` file and how to scale 3D objects in our code. We looked at the `text` operation and created a dynamic plaque that adjusted its size to the text that sat on it.

Modules gave us the ability to organize our code more effectively. By creating modules based on actions, we were able to redesign our Thumbs Up award quickly.

In the next chapter, we will investigate some of the most common OpenSCAD libraries as we continue to learn more about this powerful CAD environment.

6

Exploring Common OpenSCAD Libraries

Reusing code written by others is an excellent way to speed up our CAD design process. With OpenSCAD, there are numerous libraries of code that we can utilize to create new and exciting designs. In this chapter, we will create a desk drawer using the BOSL Standard Library. Afterward, we will re-use the code we wrote to create our own OpenSCAD library.

In this chapter, we will cover the following:

- Exploring the OpenSCAD General libraries
- Exploring the OpenSCAD Single Topic libraries
- Creating our own OpenSCAD library

Technical requirements

The following is required to complete this chapter:

- Any late-model Windows, macOS, or Linux computer that can install OpenSCAD.
- The code and images for this chapter can be found here: `https://github.com/PacktPublishing/Simplifying-3D-Printing-with-OpenSCAD/tree/main/Chapter6`.

Exploring the OpenSCAD General libraries

The General libraries in OpenSCAD includes the BOSL, dotSCAD, NopSCADlib, and BOLTS libraries. Implementing these libraries allows us to add things such as threaded rods, modeled parts (parts that are not 3D-printed but are used in designs), and mathematically complex shapes. The following section includes a short breakdown of each of these libraries.

BOSL

The **Belfry OpenSCAD Library (BOSL)** consists of operations to create shapes such as rounded boxes and threaded rods. Operations to enhance OpenSCAD's translate and rotate operations are also included in the BOSL.

In *Figure 6.1*, we can see a threaded rod created using the BOSL:

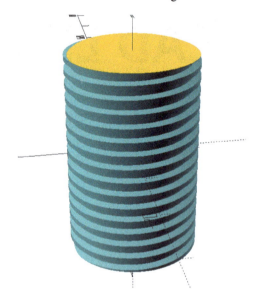

Figure 6.1 – A threaded rod created with the BOSL

We will be exploring the BOSL in more detail in the upcoming *Using the BOSL to design a desk drawer* section.

dotSCAD

The **dotSCAD** library aims to reduce mathematical complexity when using OpenSCAD. We can utilize dotSCAD to create complex shapes for our designs. In *Figure 6.2*, we can see a rose created in OpenSCAD using the dotSCAD library:

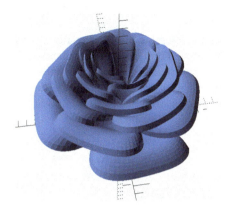

Figure 6.2 – An OpenSCAD rose using the dotSCAD library

This rose can be modified and easily put into our designs, saving us the hassle of importing a rose as a 3D object.

NopSCADlib

The **NopSCADlib** library provides modeled parts for use in our OpenSCAD designs. These parts include things such as bearings, batteries, and parts for RepRap 3D printers that cannot be 3D-printed.

In *Figure 6.3*, we can see a hygrometer rendered in OpenSCAD using the NopSCADlib library:

Figure 6.3 – A hygrometer from the NopSCADlib library rendered in OpenSCAD

This object was rendered using a single line of code and represents the standard mini hygrometer that can be purchased online at places such as Amazon and eBay.

> **Hygrometers and 3D Printing**
>
> Hygrometers measure ambient temperature and relative humidity and are useful tools for 3D printing. For hygroscopic filaments, such as nylon, it is important to keep the environment as dry as possible when storing and printing. Hygrometers placed with filaments in vacuum bags or in 3D printer enclosures allow us to measure the relative humidity and adjust the environment (such as the addition of silica packets).

Rendering non-3D-printable objects in OpenSCAD allows us to design around objects such as hygrometers. This saves us from having to measure the part in the real world and compensate for it in our OpenSCAD design.

BOLTS

BOLTS (not an acronym) is a free and open source library of standard parts that we can incorporate into our OpenSCAD designs. These parts consist mainly of standard nuts, bolts, pipes, and so on that we can use with our 3D-printed parts in our projects.

In *Figure 6.4*, we can see a pipe generated using the BOLTS library. Three parameters were used to create this pipe – the inside diameter (8 mm), the outside diameter (10 mm), and the length of the pipe (50 mm). The part was generated with the `pipe(10, 8, 50)` command:

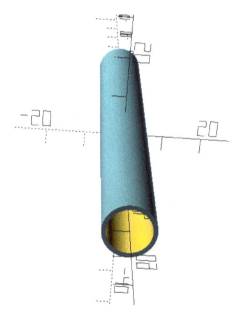

Figure 6.4 – A pipe generated with the BOLTS library in OpenSCAD

Now that we have looked briefly at the OpenSCAD standard libraries, let's create a design using what we have learned. We will design a desk drawer by utilizing the BOSL. To create this design, we will download and install the library onto our computer. We will then use modules from the BOSL to create our desk drawer – a drawer with rails that rides on sliders bolted to the underside of a desk.

Let's get started.

Using the BOSL to design a desk drawer

As mentioned in the previous section, there are shapes such as rounded boxes that we can create using the BOSL. We will design our desk drawer using these shapes from the BOSL.

We will start by downloading and installing the library.

Downloading and installing the BOSL

To download the BOSL, we do so from the OpenSCAD website. We then unzip, rename, and copy the contents into our OpenSCAD `libraries` folder.

To do so, follow these steps:

1. Navigate to the OpenSCAD libraries web page using the following URL: `http://www.openscad.org/libraries.html`.

2. Click on the **Library** link under **BOSL**. This will take us to the GitHub page for the BOSL.

3. Click on the green **Code** drop-down button and select **Download ZIP**.

4. Download and unzip the file. Observe that there is a folder called `BOSL-master`. Open this folder.

5. Observe that there is a folder with the same name (`BOSL-master`) inside. Rename this folder `BOSL`.

6. In OpenSCAD, click on **File | Show Library Folder...**. Observe that the `libraries` folder opens.

7. Copy the `BOSL` folder to the `libraries` folder.

We have now installed the BOSL into our OpenSCAD installation. Let's start our design by creating the tray portion with a BOSL rounded box.

Creating the drawer tray

Before we create our design, we must import the libraries we require and set any variables we will use. To create the drawer tray of our desk drawer, we will start with a rounded box from the `shapes.scad` BOSL file:

1. In a new OpenSCAD file, add the following line to the top of the file:

    ```
    use <BOSL/shapes.scad>
    ```

    Observe that we do not need to put a semicolon at the end of the `use` line.

2. We will now set the variables for our design. Add the following after the first line:

    ```
    $fn=200;
    width = 190;
    length = 190;
    height = 70;
    rail_size=10;
    hollow_factor = 0.95;
    ```

    We already know that `$fn` sets the resolution of the design. The rest of the variables will be used to create the objects that will form our design.

3. Next, add the following line below the variables, which will generate a cuboid shape from the BOSL:

    ```
    cuboid([width,length,height], fillet=10);
    ```

    We defined `width`, `length`, and `height` in our variable declarations. The `fillet` value sets the roundness of the cuboid.

4. Click on **Render** or hit *F6* on the keyboard (or *F5* to preview if the design is taking too long to render). Observe the following shape:

Figure 6.5 – A cuboid generated with the BOSL

5. We now need to cut our shape in half and hollow it out. We also want to place the code to generate the drawer tray in a module. Replace the `cuboid` code with the following:

```
module create_tray()
{
    difference()
    {
        cuboid([width,length,height], fillet=10);

        scale([hollow_factor,hollow_factor,hollow_
          factor])
        cuboid([width,length,height], fillet=10);

        translate([0,0,height])
        cube([width*2,length*2,height*2],
          center=true);
    }
}
create_tray();
```

There is quite a lot of code here, so let's go through it before we render. What we are essentially doing is taking the difference between our original cuboid and a scaled-down version of it. This scaled-down version is 5% smaller as the `hollow_factor` value is `0.95` (set with the initial variable declarations), and we use the `scale` operation to reduce a new version of the cuboid by this much. We then subtract a standard cube to cut the hollowed-out cuboid in half. We simply double the `width`, `length`, and `height` values on the standard cube to get a clean cut. We move the cube up by the value of `height` using the `translate` operation so that it will be above the *z* axis. We need to do this, as the cube is centered with all axes when it is created. Note that we move the cube up by the value of `height` and not `height/2` as we usually do for objects that are centered on the *z* axis. This is due to the `height*2` size of our cutaway cube.

6. Click on **Render** or hit *F6* on the keyboard (or *F5* to preview if the design is taking too long to render). Observe the following shape:

Figure 6.6 – The tray generated using create_tray() module

Now that we have created the tray for our desk drawer, it is time to add side rails and sliders to our design. This will allow us to mount our tray under a desk or table.

> **Using include or use**
>
> We can import the `shapes.scad` file into our design by using an `include` statement instead of `use`. The difference between the two is in how `shapes.scad` is implemented. Both statements will bring in all the modules from `shapes.scad`. However, the `include` operation will execute any code that sits outside of any modules in `shapes.scad`, while the `use` operation will not. The difference between the two will become clearer when we create our own library in the *Creating our own OpenSCAD library* section.

We will now turn our attention to using rails and sliders from the BOSL. We use the `sliders.scad` file for both. We start by adding rails to our drawer.

Adding rails to our drawer tray

The BOSL contains rails and sliders that we can use for our desk drawer. These parts give us the objects we need to allow our desk drawer to slide in and out from under our table or desk.

To add rails, follow these steps:

1. Add the library for rails and sliders with the following code at the top of our file:

```
use <BOSL/sliders.scad>
```

2. Comment out the `create_tray();` line so that we can focus on the rails:

```
//create_tray();
```

3. Add the following code:

```
rail(l=length-20, w=rail_size, h=rail_size);
```

What we have done here is to create a rail that is the `length` value of our drawer minus 20 mm. We make it shorter to account for the round corners of the tray. We have already declared `rail_size` to be 10 in our variable declarations, and we use this variable to define the `width` (w) and `height` (h) values of our rail.

4. Click on **Render** or hit *F6* on the keyboard (or *F5* to preview if the design is taking too long to render). Observe the following shape:

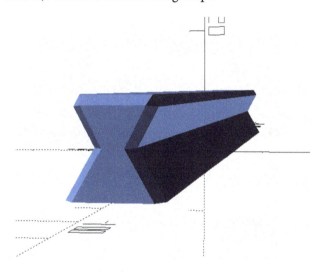

Figure 6.7 – The rail generated using the BOSL Standard Library

5. As we can see, we can use this shape for rails on our drawer. However, we must first put the shape in the right place. We want to encapsulate our code in a module as well. Delete the line `rail(l=length-20, w=rail_size, h=rail_size);` and add the following module below the `create_tray()` module:

```
module create_rail()
{
    translate([width/2,0,-(rail_size/2)])
    rotate([0,90,0])
    rail(l=length-20, w=rail_size, h=rail_size);
}
```

What we have done here is rotate the rail to a position that will make it useful. We then move it to the right by half the `width` value, as the tray is positioned in the center of our design. We move it down by half of its `height` (`rail_size/2`) value so that it will be below the *z* axis and line up with our tray.

6. Before we can see the results of our changes, we need to call `create_rail()` and uncomment out the `create_tray()` module:

```
create_tray();
create_rail();
```

7. Click on **Render** or hit *F6* on the keyboard (or *F5* to preview if the design is taking too long to render). Observe the following shape:

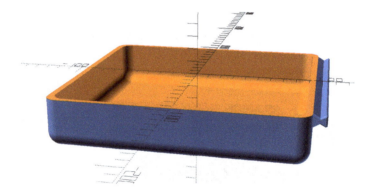

Figure 6.8 – The tray with a rail added

8. As we can see, a rail has been added to the right side of our tray. We can modify the `create_rail()` module by adding code to put a rail on the left side as well; however, there is an easier solution. Modify the non-module code (code that sits outside of a module and is run when we render) to the following:

```
create_tray();
create_rail();
mirror([1,0,0])create_rail();
```

The `mirror` operation does exactly what its name implies – it creates a mirror of an object. The parameters determine where the mirroring occurs based on *x*, *y*, and *z* values. In our case, we are mirroring in the *x* direction.

9. Click on **Render** or hit *F6* on the keyboard (or *F5* to preview if the design is taking too long to render). Observe that there are now two rails added to our tray:

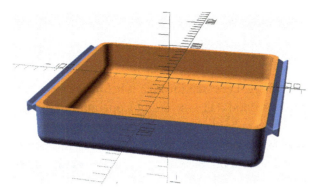

Figure 6.9 – The tray with both rails added

As we can see, the rails give our drawer something to mount inside a slider. Before we create the sliders for mounting the rails, we will add a handle to the drawer.

Creating the handle for our drawer

To create a handle, we will use cuboid from the BOSL and cut away a portion of it. We will position this handle on the front of the drawer. We will place the code to do this in a module.

To create the handle, follow these steps:

1. Create a new module below the create_rail() module with the following code:

```
module create_handle(size)
{
    translate([0,-length/2,-(size*1.5)])
    difference()
    {
        difference()
        {
            cuboid([5*size,3*size,1.5*size],
            fillet=2);

            scale([0.8,0.8,2])
            cuboid([5*size,3*size,1.5*size],
            fillet=2);
        }
        translate([0,size*50,0])
        cube([size*100,size*100,size*100],
```

```
                    center=true);
       }
   }
```

There is a lot of code here. Let's step through it before we move on to implementing it. If we start with the second `difference()` operation, we can see that we take the difference between a cuboid created to be 5 times the `size` parameter in the *x* direction, 3 times in the *y* direction, and 1.5 times in the *z* direction, with a version of the same cuboid but 80% of its size in the *x* and *y* directions and 200% in the *z* direction.

These multiplication values are arbitrary and were chosen for the shape that they create. With the first `difference()` operation, we simply cut our handle in half with a box that is much bigger and moved over to be on one side of the *y* axis. We then moved the whole shape in place with the first `translate()` operation, placing it in front of our tray.

2. To view our design so far, modify the non-module code to look like the following:

```
create_tray();
create_rail();
mirror([1,0,0])create_rail();
create_handle(10);
```

With `create_handle(10);`, we are creating a handle with a starting value of `10`. The result will be a handle that is 50 mm in the *x* direction, 30 mm in the *y* direction, and 15 mm in the *z* direction.

3. Click on **Render** or hit *F6* on the keyboard (or *F5* to preview if the design is taking too long to render). Observe that our drawer is now complete:

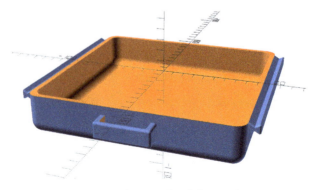

Figure 6.10 – The completed drawer design

For reference, here is a picture of a drawer that has been 3D-printed and painted:

Figure 6.11 – A 3D-printed desk drawer

Now that we have completed the drawer portion of our desk drawer, it is now time to design the sliders that allow the drawers to slide in and out under the table.

Creating the sliders for our desk drawer

As mentioned in the *Adding rails to our drawer tray* section, there are slider objects in the BOSL. We will create a slider and then move it into the correct position.

Let's get started:

1. Create a new module with the following code:

```
module create_slider(offset)
{
    base=20;

    slider(l=length,h=rail_size,base=base,
    wall=4,slop=offset);
}
```

What we are doing here is creating `slider` equal to the `length` value of the drawer with a `base` value of `20` mm. The `offset` value is used to create some space between the rail and the slider.

2. To see the slider by itself, comment out all the other non-module code and add a call to the `create_slider()` module, as shown in the following snippet:

```
//create_tray();
//create_rail();
//mirror([1,0,0])create_rail();
//create_handle(10);
create_slider(0.4);
```

We are calling the `create_slider()` module with an `offset` value of `0.4`. This value may be experimented with to provide a snug fit between the rail and the slider.

3. Click on **Render** or hit *F6* on the keyboard (or *F5* to preview if the design is taking too long to render). Observe that we see a slider that looks like the following (ensure that we have use the line `<BOSL/sliders.scad>` at the top of our code):

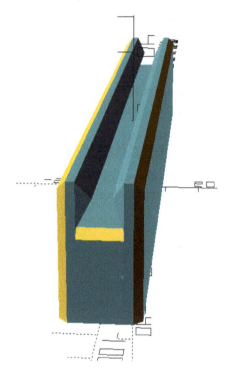

Figure 6.12 – A slider generated in OpenSCAD using the BOSL Library

4. Now, we will move the slider into place. However, please note that this step is only necessary to check the fit; this is because when we 3D-print the slider, we do so separately from the drawer, as it is not attached to it. Change the code for `create_slider()` to the following:

```
module create_slider(offset)
{
    base=20;

    translate([width/2+(rail_size+base),0,
    -rail_size/2])
    rotate([0,-90,0])
```

```
        slider(l=length,h=rail_size,base=base,
        wall=4,slop=offset);
  }
```

To understand what we just did here, let's work from the bottom to the top. With this change, we rotate `slider` `-90` degrees on the *y* axis. We then move it to the right on the *x* axis by a value that is equal to half the `width` value (as the drawer is centered) plus the size of the rail and slider `base` added together (`width/2+(rail_size+base)`). We then move the slider down so that it lines up with the rail (`-rail_size/2`).

5. To view all the parts we have so far, uncomment out the code commented in *Step 2*. Add a second `create_slider()` call and use the `mirror` operation to create a slider on the other side of the tray. Our non-module code should look like the following:

```
create_tray();
create_rail();
mirror([1,0,0])create_rail();
create_handle(10);
create_slider(0.4);
mirror([1,0,0])create_slider(0.4);
```

6. Click on **Render** or hit *F6* on the keyboard (or *F5* to preview if the design is taking too long to render). Observe that the sliders have been created:

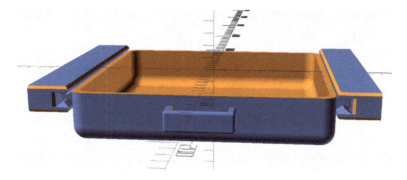

Figure 6.13 – The sliders generated and put in place

Now that we have created the sliders for our desk drawer, there is one task that remains. We need to create screw holes on the sliders so that they may be mounted onto the bottom of our desk or table.

In the next section, we will do just that.

Adding screw holes to the sliders

We will create a separate module to create the screw holes. We will then call that module from the create_slider() module. This will keep our code clean and also allow code re-use.

Let's get started:

1. Between the last module and the non-module code, create a new module with the following:

```
module create_3mm_screw_hole()
{
    union()
    {
        cylinder(d=3, h=500);

        translate([0,0,-500])
        cylinder(d=10, h=500);
    }
}
```

2. Comment out all the non-module code and put the following at the bottom:

```
//create_tray();
//create_rail();
//mirror([1,0,0])create_rail();
//create_handle(10);
//create_slider(0.4);
//mirror([1,0,0])create_slider(0.4);
create_3mm_screw_hole();
```

What we are doing here is simply creating two cylinders stacked together, one that is 3 mm in diameter and one that is 10 mm in diameter. The 3 mm hole is the actual hole our screw will go through; the 10 mm one is the countersink hole for the screw. We have made both cylinders extremely long to provide for clean cuts. The cylinders sit in the middle on the *z* axis.

3. Click on **Render** or hit *F6* on the keyboard (or *F5* to preview if the design is taking too long to render). Observe that our cylinders reach above and below the view in the object display area:

Figure 6.14 – Cylinders created from the create_3mm_screw_hole() module

4. We will now modify the code in the `create_slider()` module. Change the code in the module to the following:

```
module create_slider(offset)
{
    base=20;
    hole_inset=20;

    difference()
    {
        translate([width/2+(rail_size+base),0,
         -rail_size/2])
        rotate([0,-90,0])
        slider(l=length,  h=rail_size,
         base=base,wall=4,  slop=offset);

        translate([width/2+(rail_size+(base/2)),
         (length/2)-hole_inset,-rail_size/2])
        create_3mm_screw_hole();

        translate([width/2+(rail_size+(base/2)),
         -((length/2)-hole_inset),-rail_size/2])
        create_3mm_screw_hole();
    }
}
```

Before we render our design, let's go through the changes. Basically, what we are doing in the new `create_slider()` module is taking the difference between `slider` and a 3 mm screw hole. We have added a new variable called `hole_inset`, which is the value from the ends of the slider where we want to place our screw holes. In this case, we are putting our screw holes 20 mm from the ends. The `translate()` operation moves the holes to the base part of the slider, away from the side where the drawer will slide. The second `translate()` operation differs from the first by taking the negative value in the *y* direction to make a mirrored copy. The holes are then moved down in the *z* direction by half the `rail_size` value to countersink the screw hole.

5. Before we render our design, we will uncomment out all the non-module code and delete the last non-module line we added in *Step 2*. Our non-module code should look like the following:

```
create_tray();
create_rail();
mirror([1,0,0])create_rail();
create_handle(10);
create_slider(0.4);
mirror([1,0,0])create_slider(0.4);
```

6. Click on **Render** or hit *F6* on the keyboard (or *F5* to preview if the design is taking too long to render). Observe that our design is now complete, and the sliders now have screw holes with countersinks:

Figure 6.15 – The final design with screw holes

We have now finished our design of the desk drawer. As we can see, we are able to easily leverage the OpenSCAD standard library to create new and interesting designs.

In the next section, we will look at OpenSCAD Single Topic libraries.

Exploring OpenSCAD Single Topic libraries

Now that we understand how OpenSCAD standard libraries can be utilized to improve our designs, let's look at what are called OpenSCAD Single Topic libraries. As their name implies, OpenSCAD Single Topic libraries are used for specific purposes, such as creating a threaded nut for a design. There are six libraries listed on the OpenSCAD website (http://openscad.org/libraries.html) under **Single Topic**. We will look at the four most relevant (for our purposes) libraries:

- Round Anything
- Mark's Enclosure Helper
- The OpenSCAD threads.scad module
- The OpenSCAD smooth primitives library

Let's start with the Round Anything library.

Round Anything

The motivation behind the Round Anything library is, as its name implies, to round parts. Standard OpenSCAD code lacks the functionality for rounding basic shapes. Installing this library provides a good tool to use in our designs.

For our example, however, we will look at the shell2d() operation, which ironically does not round a part. With shell2d(), we create a hollow outline of a 2D shape. We can then use the gridpattern() operation from Round Anything to create a grid inside the new shape.

In *Figure 6.16*, we can see the effect of the shell2D() operation on a 2D sketch. The image on the left is the shape we created in *Chapter 4*, *Getting Started with OpenSCAD*, for the PVC hook. Applying the shell2D() and gridpattern() operations on this shape creates what we see on the right side:

Figure 6.16 – The before and after shell2D() and gridpattern() operations

To create the shape on the right, follow these steps:

1. Download and install the Round Anything library from this URL: `https://openscad.org/libraries.html`.

2. Create a new design with the following code:

```
use <Round-Anything/roundAnythingExamples.scad>
shell2d(0, -5)
{
    intersection()
    {
        translate([20,0])
        circle(d=80);

        translate([30,0])
        square([60, 70], center=true);
    }
    gridpattern(iter=50);
}
```

In this code, we create `square` with one rounded side using the `intersection()` operation. We then wrap that inside the `shell2d()` operation with offset values of 0 and -5. This defines the thickness of the shell (negative numbers create the shell inward, and positive numbers outward). For these numbers, we will create a shell with an inside wall thickness of 5 mm. We then add `gridpattern(iter=50);`, which creates a pattern inside our shape.

3. Click on **Render** or hit *F6* on the keyboard (or *F5* to preview if the design is taking too long to render). Observe that the pattern looks like the pattern on the right side in *Figure 6.16*.

4. To experiment further, comment out the `gridpattern()` operation:

```
//gridpattern(iter=50);
```

5. Click on **Render** or hit *F6* on the keyboard (or *F5* to preview if the design is taking too long to render). Observe that the 2D shape is a shell with a value of 5 mm from the outside edge inward:

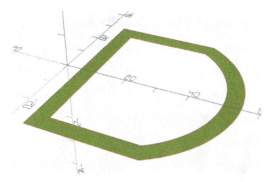

Figure 6.17 – The shape after the shell2D() operation without gridpattern()

There is much more functionality to the Round Anything library than we have touched on. What is important to gain from this exercise is the power that this external library can bring to our designs.

We will now look at a library that will build enclosures for us.

Mark's Enclosure Helper

With this library, we can easily create enclosures for our projects with a few lines of code. We can make enclosures with interlocking rims, snap-fit enclosures, and rounded corners.

For our example, we will create a simple hinged enclosure with a few lines of code. To create our hinged box, follow these steps:

1. Download and install the Mark's Enclosure Helper library from this URL: https://openscad.org/libraries.html. Be sure to rename the folder from MarksEnclosureHelper-master to MarksEnclosureHelper.

2. Create a new design with the following code:

```
include <MarksEnclosureHelper/hingebox_code.scad>
hingedbox( box_def );
hinge_points = [0.5];
hinge_len = 20;
```

What we have done here is define a standard hinged box enclosure. It may seem odd that we are simply setting variables without passing them into a module or operation, but take note of include at the top. This library relies heavily on code that is written outside of modules. If we were to replace include with use, the code will not work. The hinge_points variable defines where the hinges are located, with the 0.5 value putting them in the middle of the enclosure. The hinge_len variable defines the size of the hinge, which we set to 20.

3. Click on **Render** or hit *F6* on the keyboard (or *F5* to preview if the design is taking too long to render). Observe the enclosure created:

Figure 6.18 – The enclosure generated using the Mark's Enclosure Helper library

A common task among makers with 3D printers is creating enclosures for various projects. Having a library that does this with ease is a great tool to have.

Now that we know how to create quick enclosures, let's turn our attention to creating screws and bolts that we can use in our designs.

The OpenSCAD threads.scad module

The OpenSCAD threads.scad library is designed to be an efficient way to generate bolts, threaded rods, and nuts that we can use in our projects.

In our example, we will create a 25 mm M10 bolt with a couple of lines of code. Let's get started:

1. Download and install the OpenSCAD threads.scad library from this URL: https://openscad.org/libraries.html.

2. Create a new design with the following code:

```
use <threads.scad>
MetricBolt(10, 25, tolerance=0.4);
```

This code couldn't be simpler. We pass in 10 for the diameter and 25 for the length. We can adjust tolerance if we are finding our 3D prints of this bolt too tight or too loose.

3. Click on **Render** or hit *F6* on the keyboard (or *F5* to preview if the design is taking too long to render). Observe that an M10 25 mm bolt is generated:

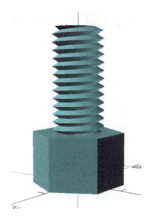

Figure 6.19 – An M10 25 mm bolt generated from the OpenSCAD threads.scad library

4. As we can see in the first line of the code in *Step 2*, the threads.scad library
    is brought into our program with the use command. If we were to change this to
    include, we would see a demo of the library. To verify this, change the code to the
    following:

```
include <threads.scad>
MetricBolt(10, 20, tolerance=0.4);
```

5. Click on **Render** or hit *F6* on the keyboard (or *F5* to preview if the design is taking
    too long to render). Observe that many parts are generated, including our
    M10 25 mm bolt:

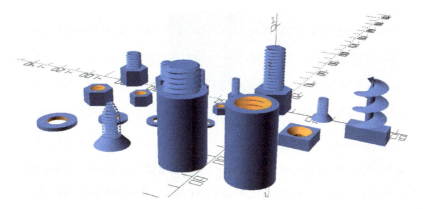

Figure 6.20 – A generated demo of parts from the OpenSCAD thread.scad library

Now that we can see there is an easy way to generate threaded nuts and bolts, let's turn our
attention to a library that generates smooth basic objects for us.

The OpenSCAD smooth primitives library

This library provides common primitive shapes with additional parameters to smooth out the shape. As many of us know, it is not always easy to round off or smooth an edge of a common shape in OpenSCAD. Although this library is not vast, it does provide a few useful shapes that we can implement in our designs.

For our example, we will look at SmoothCylinder. To explore this, follow these steps:

1. Download and install the OpenSCAD smooth primitives library from this URL: https://openscad.org/libraries.html.

2. Create a new design with the following code:

    ```
    use<smooth_prim.scad>
    SmoothCylinder(10, 30, 5);
    ```

    In the code, we are creating SmoothCylinder with a radius of 10 mm, a length of 30 mm, and a smoothing radius of 5 mm.

3. Click on **Render** or hit *F6* on the keyboard (or *F5* to preview if the design is taking too long to render). Observe that a cylinder with rounded edges is rendered:

Figure 6.21 – SmoothCylinder generated with the OpenSCAD smooth primitives library

It is easy to see how we can use SmoothCylinder in our OpenSCAD designs and how easy it is to create one. For example, by cutting SmoothCylinder in half and hollowing it out, we can create a bell jar-style cover for a project.

Now that we have a deeper understanding of how to use external libraries, let's create one of our own. We will use the code we wrote to create the desk drawer in the *Using the BOSL to design a desk drawer* section.

Creating our own OpenSCAD library

One way to turn our desk drawer code into a library file is to add it to our OpenSCAD installation's `libraries` directory. We can open this location on our computer by clicking on **File | Show Library Folder...** in OpenSCAD.

Before we do that, we should take note of how we will be using this library. If we were to simply copy the code as we left it in the *Using the BOSL to design a desk drawer* section, we could see the whole drawer with the sliders generated. This would happen when using an `include` statement for importing:

```
include <desk_drawer.scad>
```

This is due to the non-module code at the end of the file that creates the drawer and the sliders we added in the *Using the BOSL to design a desk drawer* section. We will, however, be able to change the size of the drawer and sliders, as we will have access to the non-module variables declared at the beginning of the `desk_drawer.scad` library file. This may seem like the way we should approach creating our own OpenSCAD library file. However, it would make little sense when it comes to 3D printing, as our 3D file export (`.stl`, `.3mf`) would have both the drawer and the sliders together. This would prove difficult to 3D-print as separate parts.

Another solution is to bring our library file into our new file with the `use` statement:

```
use <desk_drawer.scad>
```

With this approach, we solve the issue of the drawer and sliders rendered together, but we do not have access to the variables that would allow us to change the size of the drawer. Also, we would expect the user of this library to know which modules to call to build the various components of the drawer, such as `create_tray()` and `create_rail()`, and the corresponding `mirror` operation for creating the opposite side rail. The drawer would always be the size set by the variables at the beginning of the file unless we changed the input parameters for the modules to accept values for `length`, `width`, and `height`. This would add extra complexity for anyone using this library.

The best solution is to modify our file and instruct our user to use our library like any other OpenSCAD library by using the `include` statement.

Let's do just that:

1. Create a copy of the `desk_drawer.scad` file and call it `desk_drawer_lib.scad`.

2. Save `desk_drawer_lib.scad` in the OpenSCAD `libraries` folder.

3. In `desk_drawer_lib.scad`, remove all the non-module code from the bottom of the file. Do not remove the variable declarations from the top of the file.

4. We will now create two new modules that will make using this library easier. We will start with a module to create the drawer. Add the following to the bottom of the code:

```
module create_drawer()
{
    create_tray();
    create_rail();
    mirror([1,0,0])create_rail();
    create_handle(10);
}
```

5. To create the sliders at the side of the drawer, we will add a new module. Add the following module to the bottom:

```
module create_sliders(offset=0.4)
{
    create_slider(offset);
    mirror([1,0,0])create_slider(offset);
}
```

6. Take note of the *plural* in the name of our new module and the default value of 0.4 for `offset`. This will make the library a little easier to use, as the user will not have to remember to supply an `offset` value. For good measure, we should add a default value to the original `create_slider()` module so that it looks like the following:

```
module create_slider(offset=0.4)
{
    base=20;

    translate([width/2+(rail_size+base),0,
    -rail_size/2])
    rotate([0,-90,0])
    slider(l=length,h=rail_size,base=base,
    wall=4,slop=offset);
}
```

This new `offset` value makes the library a little easier to use for calls to `create_slider()`. We use `create_slider()` when we need to generate a single slider.

7. We are now ready to use our new library. Create a new file in OpenSCAD with the following code:

```
include <desk_drawer_lib.scad>
width=500;
create_drawer();
create_sliders();
```

With this code, we import our library, and then we set the `width` value to `500`. We then create the drawer and the sliders.

8. Click on **Render** or hit *F6* on the keyboard (or *F5* to preview if the design is taking too long to render). Observe that a new desk drawer complete with sliders is generated and that its width is greater than its length:

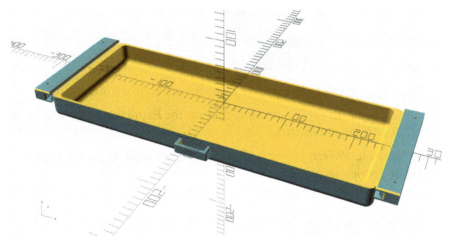

Figure 6.22 – A wide drawer generated with the custom OpenSCAD library

By arranging our library code this way, it becomes easy to generate objects for 3D printing. For example, if we wanted to print a single slider, we could generate one with this code:

```
create_slider();
```

9. Click on **Render** or hit *F6* on the keyboard (or *F5* to preview if the design is taking too long to render). Observe that a single slider is generated:

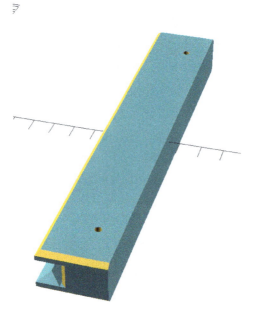

Figure 6.23 – A single slider generated with the custom OpenSCAD library

10. To generate a `.stl` file for 3D printing, click on the **Export as STL** button in the editor or hit *F7* on the keyboard.

11. Save the `.stl` file. We can view it with an STL viewer, such as 3D Viewer in Windows or Preview in macOS:

Figure 6.24 – A slider as a .stl view in 3D Viewer in Windows

With our object stored as a `.stl` file, we can then proceed to load it into a slicer program, such as Cura, and prepare it for 3D printing.

Having our desk drawer code as an installed library means we can utilize it to generate desk drawers easily. This allows us to place such a component in larger designs – for example, a design of an entire workshop.

The ability to see our designs before construction is the power of OpenSCAD and CAD design in general. It limits any measurement mistakes and, thus, any post-construction modifications.

Summary

In this chapter, we introduced external libraries to our designs. We were able to use the BOSL to design a desk drawer that slides on sliders under our desk. We also explored many of the libraries available from the OpenSCAD website, noting the design inspirations and strengths of the libraries.

We were able to then take the code written for our desk drawer design and implement it as an OpenSCAD library. As we saw, doing this greatly simplifies future designs. It allows us to design something rather complex by breaking it down into separate components.

With this chapter, we come to the end of the second part of this book, *Learning OpenSCAD*, where we explored OpenSCAD from basic to more complex concepts. We will use this knowledge in the third part of this book, *Projects*, as we use OpenSCAD to design and then bring our designs to life through 3D printing.

Part 3: Projects

Now that we understand 3D printers and 3D design, let's start to create some objects of our own. Our projects will start off with a simple 3D printed name badge before we introduce more advanced design concepts to build a riser for our laptop. We will increase the complexity by creating a model rocket from scratch, building it around a paper tube from a roll of paper towels.

In this part, we cover the following chapters:

- *Chapter 7, Creating a 3D-Printed Name Badge*
- *Chapter 8, Designing and Printing a Laptop Stand*
- *Chapter 9, Designing and Printing a Model Rocket*

7
Creating a 3D-Printed Name Badge

In *Part 1*, *Exploring 3D Printing*, we covered the basics of 3D printing. We looked at the various components that make up a 3D printer and had a brief overview of the several types of materials we can print with. In *Part 2*, *Learning OpenSCAD*, we jumped into OpenSCAD by starting out with simple concepts before looking into modules and libraries. In *Part 3*, *Projects*, we will combine what we learned from the first two parts of the book and go from OpenSCAD design to 3D-printed objects.

Part 3, *Projects*, starts with this chapter, where we will create a 3D-printed name badge to be worn at conventions or inside shops. We will design the name badge in OpenSCAD before importing our design into a slicer program. We will then proceed to print out our design.

In this chapter, we will cover the following topics:

- Creating text for our 3D-printed name badge
- Adding a base plate to our 3D-printed name badge
- Printing out our 3D-printed name badge

Technical requirements

The following is required to complete the chapter:

- Any late-model Windows, macOS, or Linux computer that can install OpenSCAD and Cura.

- 3D printer—any **fused deposition modeling** (**FDM**) printer should work; however, the Creality Ender 3 V2 is the printer used for our example.

- Epoxy glue.

- Brooch bar pin, as depicted in the following picture, for attaching our 3D-printed name badge to a shirt or jacket:

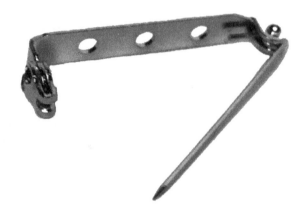

Figure 7.1 – Brooch bar pin

The code and images for this chapter can be found here:

```
https://github.com/PacktPublishing/Simplifying-3D-Printing-
with-OpenSCAD/tree/main/Chapter7
```

Creating text for our 3D-printed name badge

In *Chapter 5*, *Using Advanced Operations of OpenSCAD*, we extruded text for our *Thumbs Up Award* design. Our focus was on creating a dynamic backing plate based on the size of the text, thus we limited our exposure of OpenSCAD's text operation to a monospaced font with default alignment and size settings.

In this section, we will look more closely at the text operation and at ways that would make our text curve around shapes. Let's start by taking a closer look at the text operation.

Understanding the OpenSCAD text operation

At the time of writing, the OpenSCAD `text` operation has 10 parameters that can be set. These include parameters for font, vertical and horizontal alignment, as well as the size of text and the spacing between letters. A full list of these parameters can be found here: `https://en.wikibooks.org/wiki/OpenSCAD_User_Manual/Text`.

In this section, we will look at using a specialized font, letter spacing, font size, and text direction.

Using a specialized font

Specialized fonts give us the opportunity to add a little something extra to our designs. We may find many such fonts installed on our computer already; however, an interesting font that many of us may not have installed is the Nasalization font made by Raymond Larabie.

As its name implies, the Nasalization font is a set of characters that resemble the **National Aeronautics and Space Administration** (**NASA**) "worm" logo from the 1970s. In this section, we will download the font and test it out in OpenSCAD.

The Worm versus the Meatball

The original NASA logo featured two planets and a red chevron and was affectionately called the "meatball" logo. Considered too cluttered by the 1970s (and with a little push from a Nixon-era design improvement program), a more simplified "worm" logo was developed. However, not everyone at NASA liked the new design, and thus it was retired in 1992. The "worm", however, made a comeback and was featured in a SpaceX Falcon 9 launch.

We will start by downloading and unzipping the font, as follows:

1. Navigate to the following website and click on the **Download** link on the right-hand side: `https://www.dafont.com/nasalization.font`.

2. We will install the font in our OpenSCAD libraries folder. In OpenSCAD, click on **File | Show Library Folder...** to open the OpenSCAD libraries folder.

3. Create a new folder inside the libraries folder and call it `fonts`.

4. Unzip the `nasalization.zip` file.

5. Copy the `nasalization-rg.otf` file into the `fonts` folder.

6. Create a new OpenSCAD file and type the following code into the editor:

```
use <fonts/nasalization-rg.otf>
text("NASA", spacing=1.5,
font="Nasalization:style=Regular");
```

What we are doing here is loading the Nasalization font into our program with the `use` command. We then create text using the `text` operation, deploying the `Nasalization` font with a `Regular` style. We also set `spacing` to `1.5` to give our design a less cluttered look.

7. Click on **Preview** or hit *F5* on the keyboard. Observe here that we have recreated the NASA "worm" logo:

Figure 7.2 – NASA "worm" logo made with Nasalization font and OpenSCAD

Now that we know how to add a specialized font to OpenSCAD, let's investigate a few text attributes we may modify.

Changing the size and direction of text

A couple more parameters that we can modify with the `text` operation include the `size` and `direction` parameters. As their names imply, `size` controls the size of the text, and `direction` controls the direction.

Let's explore these parameters by doing the following:

1. Modify the code so that it looks like this:

    ```
    use <fonts/nasalization-rg.otf>
    text("NASA", direction="ttb", size=100,
    font="Nasalization:style=Regular");
    ```

    Let's start with `direction`. The values we may choose for direction are the default `ltr` (left-to-right), `rtl` (right-to-left), `ttb` (top-to-bottom), and `btt` (bottom-to-top). We are setting this value to `ttb`, which will create vertical text. We are also setting `size` to `100`, which is 10 times the default.

2. Click on **Preview** or hit *F5* on the keyboard. Observe here that the direction of the text, as well as its size, has changed:

Figure 7.3 – Vertical NASA "worm" logo made with Nasalization font and OpenSCAD

Now that we have a better understanding of the text operation, let's look at how we can curve text in OpenSCAD.

Making text curve in OpenSCAD

Bending or curving text in an arch is a common effect used in design. Bending or curving text may be used to draw attention to a part of our design, such as a name. In this section, we will create a module to bend text in a circular pattern. We will start off by learning how to analyze each letter in a text string.

Finding the first letter in a string

For text strings in OpenSCAD, we can access each letter using an index. We use open and closed square brackets to indicate an index. To test this out, do the following:

1. Create a new OpenSCAD file and type in the following code:

    ```
    test_text = "Hello OpenSCAD";
    echo(test_text[0]);
    ```

    As with many programming languages, OpenSCAD uses zero as the first index position. With our code, we are creating a string called Hello OpenSCAD and are echoing out the first letter using the [0] index.

2. Click on **Preview** or hit *F5* on the keyboard. Observe here that the letter H is shown in the console:

```
Compiling design (CSG Tree generation)...
ECHO: "H"
Compiling design (CSG Products generation)...
```

Figure 7.4 – The first letter of "Hello OpenSCAD" echoed in the console

Now that we understand how to find an individual letter from a text string in OpenSCAD, let's now explore how to iterate so that we can cycle through the text.

Using a for loop to cycle through text

As with many programming languages, OpenSCAD offers functionality for iteration in the form of a `for` loop. Using a `for` loop, we can cycle through the letters of a text string. Let's learn how to do this with a simple example, as follows:

1. Create a new OpenSCAD file and type in the following code:

```
test_text = "Hello OpenSCAD";
for(i=[0:len(test_text)-1])
{
    echo(test_text[i]);
}
```

Before we run this code, let's examine it closely. We start off by creating a string called `test_text` and setting it to the value `Hello OpenSCAD`. In the next line, we set a `for` loop to count from 0 to a value equal to the length of the `test_text` string minus 1 and assign this value to the `i` variable for every iteration through the loop. For our example, the iteration would happen 14 times. In the loop, we use `echo` to print the letter at the `i` position of `test_text` to the console.

2. Click on **Preview** or hit *F5* on the keyboard. Observe that each letter of `test_text` is printed out one by one in the console:

```
Compiling design (CSG Tree generation)...
ECHO: "H"
ECHO: "e"
ECHO: "l"
ECHO: "l"
ECHO: "o"
ECHO: " "
ECHO: "O"
ECHO: "p"
ECHO: "e"
ECHO: "n"
ECHO: "S"
ECHO: "C"
ECHO: "A"
ECHO: "D"
Compiling design (CSG Products generation)...
Geometries in cache: 37
Geometry cache size in bytes: 449344
CGAL Polyhedrons in cache: 0
CGAL cache size in bytes: 0
Compiling design (CSG Products normalization)...
Normalized tree has 1 elements!
Compile and preview finished.
Total rendering time: 0:00:00.071
```

Figure 7.5 – Letters echoed to the console

Why Do We Subtract 1 from the Length of test_text?

With many programming languages, when iterating through a list or string, we start at index 0 and stop at the value that is one less than the length. We do this using a less-than (<) operator in the for statement. This makes sense with zero-based indexes as the value of the length of an array or string—or, in our case, the number of characters in the text—is not the same as the last index value. This is due to us starting at 0 for the first index and not 1. With OpenSCAD, we do not have the option of using a less-than (<) operator in our for loop. Thus, we must subtract 1 from the length of test_text so that we can stop our iteration on the last letter of test_text.

Now that we understand how to iterate through a string of text in OpenSCAD, let's move on to rotating the text by positioning each letter individually. We will create a module to do just that.

Creating a rotate_text() module

To rotate a string of text in OpenSCAD, we will create a module to position each character based on the parameters we pass in. We will name the new module `rotate_text()` and experiment with the parameters before creating a text section for our name badge.

Let's get started, as follows:

1. Create a new OpenSCAD file and type in the following code:

```
module rotate_text(display_text,
    text_size=10,
    distance=20,
    rotation_value=360,
    tilt=0)
{
    rotate([0,0,tilt])
    for(i=[0:len(display_text)-1])
    {
        rotate([0,0,- i*rotation_value/
          len(display_text)])
        translate([0,distance,0])
        text(display_text[i],
        font="Impact:style=Regular",
        size=text_size,
        halign="center");
    }
}
```

This module takes in five parameters, with four having default values. We will start our analysis of the `rotate_text()` module by passing in only the first parameter.

2. Type the following code below the module code:

```
rotate_text("HelloOpenSCAD");
```

What we are doing here is passing the `HelloOpenSCAD` string into `rotate_text()` for the value of `display_text`. We are leaving the other four parameters at their default settings. We leave out spacing in our string to create the desired effect.

3. Click on **Preview** or hit *F5* on the keyboard. Observe here the circle of text created:

Figure 7.6 – Circle of text created using the rotate_text() module

Our code uses the Impact font, which should be found on most operating systems (feel free to use the Nasalization font from the previous section). As we can see, using default parameters gives us a circle of text, which starts at the zero position of the *y* axis. To fully utilize the `rotate_text()` module, let's look at the `text_size`, `distance`, `rotation_value`, and `tilt` parameters.

Modifying the default parameters in the rotate_text() module

By modifying the default parameters, we can create many different designs with the `rotate_text()` module. We will start by modifying the `distance` parameter. To do this, we proceed as follows:

1. Modify the code that calls `rotate_text()` so that it looks like this:

    ```
    rotate_text("HelloOpenSCAD", distance=60);
    ```

    The `distance` parameter sets the distance from the center and, thus, the radius of the invisible circle that the text is wrapped around. With the value of `60`, we triple the size of the default parameter.

2. Click on **Preview** or hit *F5* on the keyboard. Observe here that the text has spread out:

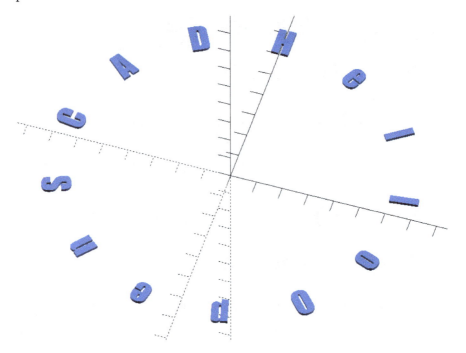

Figure 7.7 – Modifying the distance property for rotate_text()

3. We will now modify the `text_size` parameter to change the size of the text. Modify the code so that it looks like this:

```
rotate_text("HelloOpenSCAD", text_size=15, distance=60);
```

4. Click on **Preview** or hit *F5* on the keyboard. Observe here that the text is larger:

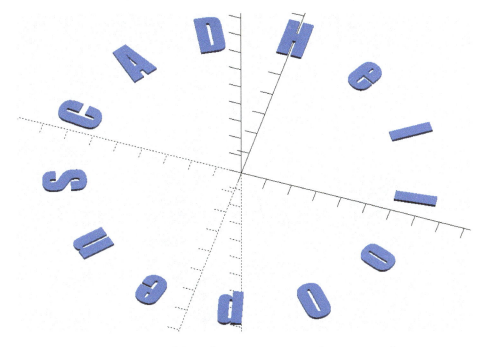

Figure 7.8 – Modifying the text_size property for rotate_text()

5. The `rotation_value` parameter defaults to `360`, which creates a full circle of text. Changing this value to anything less than `360` will create an arch instead of a circle. Modify the code so that it looks like this:

```
rotate_text("HelloOpenSCAD", 15, 60, 180);
```

We can take away the parameter names as we are setting the parameters in order. The value of `180` should create a half-circle arch of our text.

6. Click on **Preview** or hit *F5* on the keyboard. Observe here that we have indeed created a half-circle arch of our text:

Figure 7.9 – Arch of text using the rotate_text() module

7. The final parameter is `tilt`. This parameter sets the starting point of our text and defaults to `0`, which is the `0` value on the *x* axis. Modify the code to add the `tilt` parameter, as follows:

```
rotate_text("HelloOpenSCAD", 15, 60, 180, 83);
```

The value of `83` was determined by trial and error. This value may be different depending on the font and other factors.

8. Click on **Preview** or hit *F5* on the keyboard. Observe here that our arch of text has been rotated to the left:

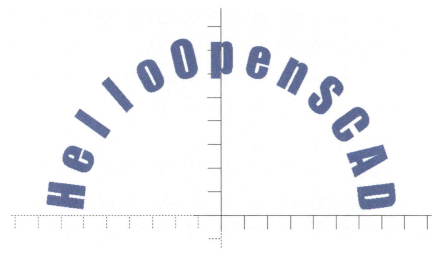

Figure 7.10 – Arch of text rotated

Now that we have a module for rotating text, it's time to put it to use to create arched text for our 3D-printed name badge project.

Creating a name tag text generator module

In this section, we will create the text portion of the 3D-printed name badge with the company name *Packt Pub* (shameless plug). We will create a new module to do this by doing the following:

1. Using the code from the *Making text curve in OpenSCAD* section, add the following module after the rotate_text() module:

```
module create_packt_name_tag_text(
    name,
    name_size=20,
    scale_factor=1)
{
    scale([scale_factor,
           scale_factor,
           scale_factor])
    union()
    {
        rotate_text("PACKT", 10, 30, 75, 30);
        rotate_text("PUB", 10, -40, -75, -23);
        text(name, size=name_size,
```

```
            font="Impact:style=Regular",
            halign="center",
            valign="center");
    }
}
```

What this code does is create the text part of a name tag for the company *Packt Pub*. The word PACKT is rotated over the top of the name. The word PUB is rotated below the name by virtue of a negative value for distance, rotational value, and tilt in the call to the `rotate_text()` module.

2. Before we can run this module, we must make a call to it in our code. Delete all non-module code (code that sits outside of modules) and type in the following code:

```
create_packt_name_tag_text("Bob Writer");
```

The `create_packt_name_tag_text()` module takes in three parameters—the `name` value to be displayed, the size of the text of the name (`name_size`), and a scale factor (`scale_factor`) to adjust the overall size of the name tag text. As `name_size` and `scale_factor` have default values, we only need to pass in the `name` parameter. We are passing in the name `Bob Writer`. Feel free to pass in your own name.

3. Click on **Preview** or hit *F5* on the keyboard. Observe the following:

Figure 7.11 – Packt Pub name badge text

You might have noticed that the name **Bob Writer** is perfectly centered with the arched text on the top and bottom. This is due to the setting of `halign` (horizontal alignment) and `valign` (vertical alignment) to `center`. Parameters for `rotate_text()` and `font` may be modified to achieve the desired effect.

4. As we will be using this code to build our 3D-printed name badge, we should save it to our OpenSCAD libraries folder. Click on **File | Save As ...** and save the file as `name-badge-text.scad` in the OpenSCAD libraries folder (**File | Show Library Folder...**).

Creating customized text is the first part of making our 3D-printed name badge. As we can see, with a few simple modules, we are able to bend text to make our design more appealing. The parameter values used have been derived by trial and error. By parametrizing our code, we can make modifications to suit any company or employee name.

In the next section, we will move on to creating a base for our 3D-printed name badge.

Adding a base plate to our 3D-printed name badge

To complete our 3D-printed name badge, we require a base plate for the text. We will start with a basic shape before we implement the code to build a base plate with a series of modules. We will then add our text to complete the design of the 3D-printed name badge. Let's start with a module to create a basic 2D shape.

Creating our first shape

We will start off the design of our base plate with a simple 2D design. As with the code we covered in the *Creating our PVC pipe hook* section of *Chapter 4*, *Getting Started with OpenSCAD*, we take the intersection of a circle and square to give us a basic first shape. We will put this code inside a module. Let's get started, as follows:

1. Create a new OpenSCAD file and type in the following code:

```
module create_first_shape()
{
    intersection()
    {
        translate([20,0])
        circle(d=85);
```

```
        translate([30,0])
        square([60, 70],center=true);
    }
}
create_first_shape();
```

The code to create our first shape is wrapped up in the `create_first_shape()` module. The `translate` values push the shape to one side of the *x* axis. The values used to define the diameter of the circle and the size of the square may be modified to create a desired effect. The module is called with this line of code: `create_first_shape();`.

2. Click on **Preview** or hit *F5* on the keyboard. Observe here that a basic shape is created:

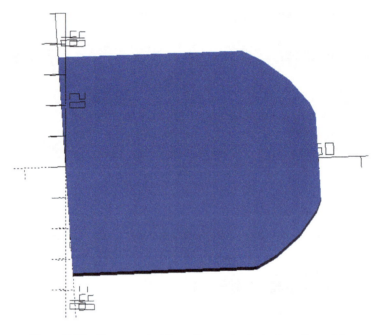

Figure 7.12 – Shape created from the create_first_shape() module

As we can see, our first shape sits on one side of the *x* axis. This sets up the shape to be mirrored later. Before we do that, we will create a module to represent the brooch pin for our 3D-printed name badge. Let's get started.

Adding an indent for the brooch pin

As we will be gluing the brooch pin to the back of the plate, a good thing to assist us is an indent or a pocket in which we can set the pin. A 1 mm indent on the back of the 3D-printed name badge should be sufficient to help us position and glue the brooch pin in place. We will create a module to do this.

Due to the different sizes of brooch pins, we will accept parameters in our module. For the brooch pin shown in *Figure 7.1*, the width is 32 mm and the height is 5 mm. Let's start by adding code for the new module, as follows:

1. Below the `create_first_shape()` module, add the following code:

    ```
    module create_brooch_indent(width, height)
    {
        translate([0,0,-1])
        color("#dc143c")
        linear_extrude(2)
        square([width, height], center=true);
    }
    ```

    With this code, we create a module called `create_brooch_indent()`, in which we take in `width` and `height` parameters. We use these parameters to create a square, which is extruded to 2 mm, colored red (`#dc143c`), and moved down in the *z* axis by 1 mm. We add color to highlight the indented region and move it down 1 mm to get a clean cut. Please note that color added to a shape only shows up in preview mode (F5).

2. Comment out the `create_first_shape();` code and add the following code:

    ```
    create_brooch_indent(32, 5);
    ```

    We are calling the `create_brooch_indent()` module with the values of 32 mm for the `width` value and 5 mm for the `height` value.

3. Click on **Preview** or hit *F5* on the keyboard. Observe here that a red shape representing the brooch pin is created:

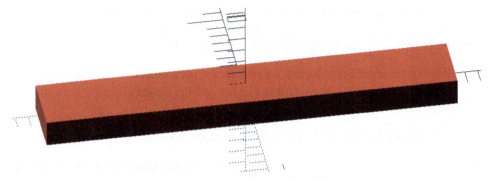

Figure 7.13 – Brooch pin indent shape

With the `create_first_shape()` and `create_brooch_indent()` modules written, it's now time to create a module that will put all the code we've written together, to make the base plate for our 3D-printed nameplate.

Putting the first shape and indent together

We will start by wrapping the `create_first_shape()` and `create_brooch_indent()` modules in a new module called `create_base_plate()`. We will then combine the name tag text with the base plate to finish our design before exporting it to a `.stl` file for 3D printing.

Coding the create_base_plate() module

We will code the `create_base_plate()` module using the `create_first_shape()` and `create_brooch_indent()` modules. With this new module, we will have all the code needed to create a base plate for our 3D-printed name badge. To create the `create_base_plate()` module, we do the following:

1. Below the `create_brooch_indent()` module, add the following code:

```
module create_base_plate(
    thickness,
    scale_factor=1)
{
```

```
    difference()
    {
        linear_extrude(thickness)
        scale([scale_factor,
               scale_factor,
               scale_factor])
        union()
        {
            create_first_shape();
            mirror([1,0,0])create_first_shape();
            circle(d=90);
        }
        create_brooch_indent(32, 5);
    }
}
```

At the heart of the `create_base_plate()` module is the union of two calls—the first one is to the `create_first_shape()` module, with the second one modified by a `mirror` operation. A circle with a diameter of 90 is added. A `scale` operation controls the size of the base plate. Of note is the absence of the `create_brooch_indent()` module from the `scale` operation as the size of the brooch pin is a set size.

2. To call the `create_base_plate()` module, remove all non-module related code and add the following code:

```
create_base_plate(2.5, 0.7);
```

The `create_base_plate()` module takes in two parameters—`thickness` (thickness of the base plate) and `scale_factor`. We pass in a value of 2.5 for the `thickness` parameter, and even though `scale_factor` has a default parameter of 1, we pass in a value of 0.7 to make the ratio of the base plate to the brooch pin smaller.

3. Click on **Preview** or hit *F5* on the keyboard. Rotate the resulting shape to view the indent for the brooch pin, as illustrated here:

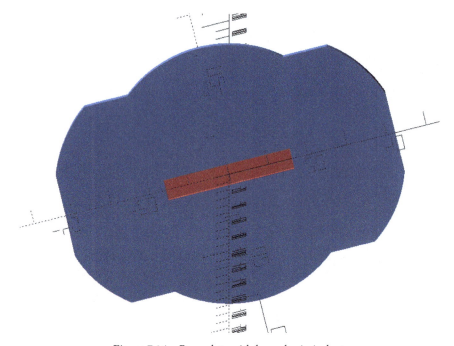

Figure 7.14 – Base plate with brooch pin indent

4. As we did with the `name-badge-text.scad` file, we will add this code to our library. Click on **File | Save As ...** and save the file as `base-plate.scad` in the OpenSCAD libraries folder (**File | Show Library Folder...**).

With our base plate code saved to the OpenSCAD libraries folder, creating our 3D-printed name badge simply involves importing `name-badge-text.scad` and `base-plate.scad` into a new file and utilizing the modules these files provide us.

Let's do that now.

Combining the text with the base plate

To finalize our design, we will create a new file and import the libraries for generating the name tag text and the base plate. To do this, we proceed as follows:

1. Create a new OpenSCAD file and type in the following code:

```
use<base-plate.scad>
use<name-badge-text.scad>
```

```
color("blue")
create_base_plate(2.5, 0.7);

color("gold")
translate([0,0,2])
linear_extrude(2)
create_packt_name_tag_text("Bob Writer",
scale_factor = 0.65);
```

In our code, we import the libraries to create a base plate and name tag text with the `use` keyword. We then create a base plate with a thickness of 2.5 mm and at a scale of 70%. The base plate is colored blue for effect. We then create a *Packt Pub* name tag using the `create_packt_name_tag_text()` module for the name Bob Writer, at a scale of 65%. Notice how we must specify the parameter name for `scale_factor` as we are not listing the parameters in order. The text is colored gold for effect, moved up 2 mm in the *z* direction, and extruded to 2 mm. If we did not move the text up, then the brooch pin indent would be covered up.

2. Click on **Preview** or hit *F5* on the keyboard. Observe here that a *Packt Pub* name badge is displayed for the name **Bob Writer**:

Figure 7.15 – Packt Pub name badge for Bob Writer

With the design finalized, it's now time to export it to a .stl file to be used for 3D printing.

Creating a file for 3D printing

In this chapter, we have been previewing instead of rendering while creating our design. To export our design as a .stl file, we need to render it. To do so, proceed as follows:

1. Click on **Render** or hit *F6* on the keyboard. Observe here that after a short bit of time, our design is rendered and is in one color:

Figure 7.16 – Rendered Packt Pub name badge for Bob Writer

2. To make use of our design in a slicer, we will export it as a .stl file. Click on **File | Export | Export as STL…** or hit *F7* on the keyboard. Give the file a name and save it to a location to be accessed by the slicer program.

It's now time to 3D print our design. To do so, we will load the .stl file into Cura and adjust the settings accordingly.

Printing out our 3D-printed name badge

To print out our design, we will open the .stl file in Cura and adjust the settings accordingly. We will consider the material, the bed adhesion, and other settings. We will then generate G-code for the print job before we store it on a microSD card and load it into our Ender 3 V2 3D printer.

We will start with loading our design and configuring the settings in Cura.

Preparing our design for a print job

We will print our name badge with two distinct colors as we want the name to stand out. This will require us to pause the print job and change out one color of **polylactic acid (PLA)** for another. We will use support for the brooch pin indent (pocket).

Before we can do all that, however, we need to load our .stl file into Cura. Let's do just that.

Loading a file into Cura

Loading a file into Cura is simple. Cura supports many different file formats, including .jpg and .png, for creating 3D prints from pictures. What we are interested in is the .stl file created from our design. To open our .stl file in Cura, we do the following:

1. In Cura, click on **File | Open File(s)...** from the top menu.

2. Navigate to the folder where the .stl file from our design is stored and click on the **Open** button.

   Observe here that our design sits centered on the build plate in Cura:

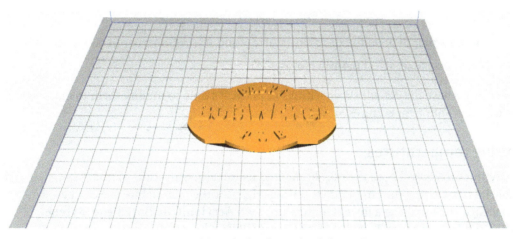

Figure 7.17 – Name badge design loaded into Cura

Now that we have our design loaded into Cura, it's time to modify the settings. We will start by loading a default slicer profile.

Using default profiles

For our projects, we will start with default templates for our slicer settings and then make modifications as needed. To select the default PLA setting in Cura, do the following:

1. In the top-middle section of the screen, click on the drop-down arrow to expand the dialog, as illustrated in the following screenshot:

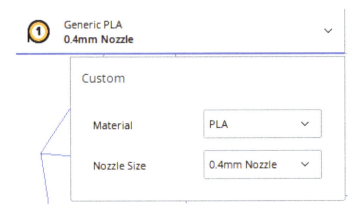

Figure 7.18 – Selecting generic slicer templates

2. For **Material**, click on the drop-down arrow and select **Generic | PLA**.

3. For **Nozzle Size**, select **0.4mm Nozzle**.

Now that we have the default slicer settings in place, it's time to modify them to our needs. We will start with temperatures.

Setting temperatures

The settings to control both the nozzle and bed temperatures in Cura sit under the **Material** tab. Getting the temperature settings right for the material we use can be a challenge. PLA melts at a lower temperature than **Acrylonitrile Butadiene Styrene** (**ABS**) and **Polyethylene Terephthalate Glycol** (**PETG**) but the temperature may vary between manufacturers (for information on the various materials for 3D printing, please refer to the *Materials available for 3D printing* section of *Chapter 1, Getting Started with 3D Printing*). Sticking with the same brand of filament is a good practice. We will be using two distinct colors of PLA from the same manufacturer in our example.

To set temperatures, we do the following:

1. Click on the down arrow on the top right of the screen to display the **Print** settings.

2. Expand the **Material** section by clicking on the down arrow of the section.

3. Set the temperatures to the following values:

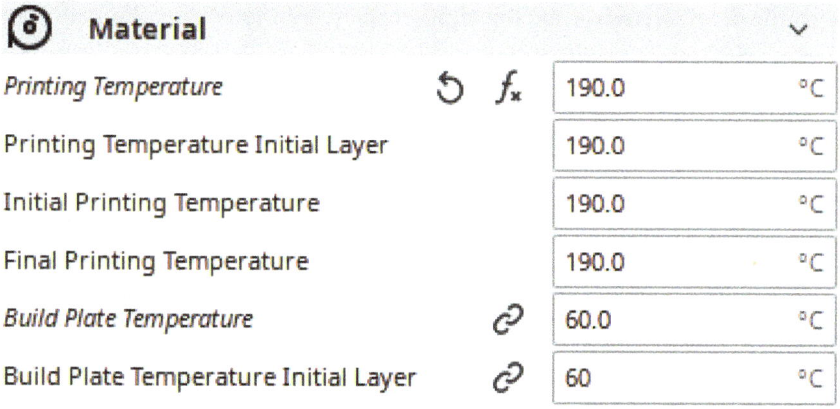

Material

Printing Temperature	↺ *f*ₓ	190.0	°C
Printing Temperature Initial Layer		190.0	°C
Initial Printing Temperature		190.0	°C
Final Printing Temperature		190.0	°C
Build Plate Temperature	⮌	60.0	°C
Build Plate Temperature Initial Layer	⮌	60	°C

Figure 7.19 – Temperature settings for the Material section

These settings may be modified based on the brand of PLA used. Generally, good quality can be achieved by keeping the temperature of the nozzle at the lower end to avoid the melted look that high nozzle temperatures may cause. However, care must be taken that the temperature is not too low so as to cause a jam.

For our example, our hot end uses the **Polytetrafluoroethylene (PTFE)** tube-to-nozzle design that comes standard with the Creality Ender 3 V2. This design works well for our purposes as PLA tends to jam when the PTFE tube does not extend directly to the nozzle. (Refer to the *Upgrading the Ender 3* section of *Chapter 1, Getting Started with 3D Printing* for more information on PTFE tube-to-nozzle design). For PTFE tube-to-nozzle designs, it is a good practice to keep temperatures on the lower side to avoid melting the PTFE tube, which generally melts at the 230°C mark. Not only does a melted PTFE tube cause jams, but it may also release hazardous toxic fumes. Our nozzle temperature of 190°C allows us to avoid such issues.

We set the bed temperature to 60°C. This should be sufficient to hold the PLA on the build plate. A too-high bed temperature softens PLA and may make removing any support material from our print difficult.

With the temperatures set, it's time to add the support.

Adding support material to our print job

The indent or pocket we use for the brooch pin of our design requires support material. This is due to the layout of our object on the build plate. Although we could get away without support material, the indent would not be as clean as it could be, thus making it harder to slot the brooch pin into place.

To add support material to our print job, we do the following:

1. Click on the down arrow on the top right of the screen to display the **Print** settings.

2. Expand the **Support** section by clicking on the down arrow of the section.

3. Check the **Generate Support** checkbox and select **Normal** for **Support Structure**, as illustrated in the following screenshot:

Figure 7.20 – Support section settings

Leave all the other settings at their defaults.

With support settings taken care of, it's time to move to the settings that will determine how our print sticks to the build plate during printing.

Adjusting the Build Plate Adhesion settings

One of the hardest things to get right with 3D printing is bed (build plate) adhesion or getting the prints to stick to the bed when we need them to stick and have them come off the bed when we need to remove them. Bed temperature is arguably the biggest factor in bed adhesion. Even though PLA may stick to the bed at room temperature, bringing up the bed temperature softens the PLA, which creates more adhesion.

Another factor that affects bed adhesion is the **Build Plate Adhesion** settings. The settings available with **Build Plate Adhesion** are **Skirt**, **Brim**, **Raft**, and **None**, with all settings except for **None** using extra filament. Ideally, we would only need to use the **None** setting, as the filament would extrude perfectly at the start of the print job and our print would stick to the build plate as we need it to.

Having the filament extrude a little bit through the nozzle before printing our object is a good practice, as any filament stuck to the nozzle from a previous print will be deposited on the build plate before printing the object. Also, freshly loaded filament may require a bit of extrusion before flowing nicely through the nozzle prior to printing the object. Extruding filament before printing an object "primes" the nozzle, giving the extrusion a running start. All the settings except for **None** may be used for this.

For more information on **Build Plate Adhesion** settings, refer to the *Build Plate Adhesion settings* section of *Chapter 3, Printing Our First Object*.

For our print, we will use **Skirt**, as we want a good flow for the extrusion but do not require extra surface area on the build plate that **Raft** and **Brim** provide due to the shape of our object (short and flat).

To set **Build Plate Adhesion**, do the following:

1. Expand the **Build Plate Adhesion** section by clicking on the down arrow of the section.

2. Set **Build Plate Adhesion Type** to **Skirt**, as illustrated in the following screenshot, and leave the other values at their default settings:

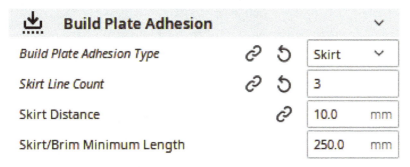

Figure 7.21 – Build Plate Adhesion settings

With our **Build Plate Adhesion** settings out of the way, it's time to look at postprocessing. With postprocessing, we will be able to pause our print so that we can swap out the filament for one of a different color.

Adding Pause at height postprocessing

Post-processing allows us to modify G-code generated for our print job. Functionality such as triggering a camera for time-lapse pictures may be implemented using postprocessing. For our purposes, we will use postprocessing to pause our print job so that we can change the filament for one with a different color. This will give us a 3D-printed name badge with text that is a different color than the base plate.

To set **Pause at height** postprocessing, we do the following:

1. Click on **Extensions | Post Processing | Modify G-Code** to access the **Post Processing Plugin** page.

2. Click on the **Add a script** button.

3. Select **Pause at height** from the list.

4. Set the values to the following:

Figure 7.22 – Pause at height settings

With these settings, the print job will stop at the height of 3 mm, where the print head is then parked at 190 mm in the *x* direction and 190 mm in the *y* direction. The nozzle temperature is set to 200°C. We add 10 degrees to soften up the PLA a little bit further for easier removal.

We will now turn our attention to slicing our object into G-code and storing it onto a microSD card to run on our printer.

Slicing our object

With our object loaded, our settings modified, and our postprocessing inserted, it's time to slice our object and store the G-code generated onto a microSD card.

To do this, we do the following:

1. Insert a microSD card into the slot on the computer. A USB or SD to microSD adapter may be required.

2. In Cura, observe the **Slice** button at the bottom-right side of the screen. The box to the left with **1** circled in red in the following screenshot indicates we are running one postprocessing event. Click on the **Slice** button:

Figure 7.23 – Cura Slice button

3. Observe that the time for the print job and the amount of material needed are displayed. As well, observe in the following screenshot that the **Slice** button turns into a button where we can select the location to store the G-code generated. Select **Save to Removable...** to save the G-code to the microSD card:

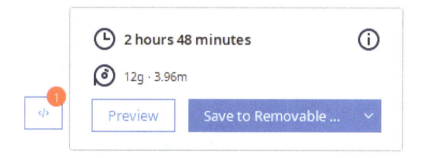

Figure 7.24 – Print job statistics and button to save G-code to microSD card

With the G-code saved onto the microSD card, it's now time to print out our object. Refer to *Chapter 3*, *Printing Our First Object* to go through the steps to do this.

Printing and finishing

The print job will run until the *z* axis reaches 3 mm, as set up in the postprocessing. When it reaches this point, we do the following:

1. Observe that the print job is paused, and a message indicating such is displayed on the screen.

2. Observe that the print head has moved to the top right of the build plate and the temperature of the nozzle is 200°C. For our example here, we use a **polyetherimide (PEI)** build plate:

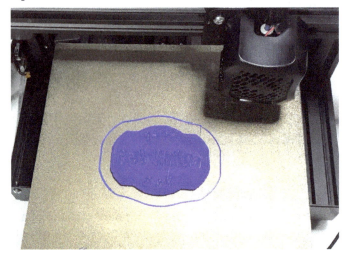

Figure 7.25 – Paused print

3. Using *Chapter 3*, *Printing Our First Object* as a reference, change the filament to a filament of a different color.

4. Click on the control knob to continue the print.

    After the print job has finished, observe that our name badge has been created, with the letters having a different color than the base plate, as illustrated in the following picture:

Figure 7.26 – Finished print

As we can see, the clarity of the letters leaves much to be desired. This may be improved by using a 0.2 mm nozzle as opposed to a 0.4 mm nozzle. Our settings in Cura would have to change to reflect this.

There is one thing left to do to complete our 3D-printed name badge, and that is to add a brooch pin to the back. This can be done easily using epoxy glue, as shown here:

Figure 7.27 – Completed 3D-printed name badge

We have created our first OpenSCAD designed object. Now that we have this experience behind us, in the next chapters, we will implement a little more complexity in our designs.

Summary

We started this chapter by looking more closely at the OpenSCAD `text` operation. We learned how to install and use a customized font as we recreated the NASA "worm" logo. From there, we explored how to change the size and direction of our text.

We implemented a `for` loop to cycle through custom text so that we could curve it around an invisible circle. We created a specialized module to do this.

From there, we created code to make a customized 3D-printed name badge using the curved text module and a module to create a base plate from a simple 2D design. We accounted for a brooch pin by providing an indent at the back of our name badge.

We proceeded to use Cura to slice our design into G-code, which we loaded onto our 3D printer. This G-code produced a print job with a pause, to allow us to change the color of the PLA for effect.

In the coming chapters, we will expand our knowledge of 3D design in OpenSCAD further as we build more complex objects. We will build a stand for a laptop in the next chapter.

8
Designing and Printing a Laptop Stand

Laptop risers allow computer users to position their laptops higher on their desks, making their desk space more ergonomic. This is usually accompanied by an external keyboard, mouse, and screen, essentially turning a laptop into a dual monitor desktop computer.

Although there are many laptop risers on the market, it would be nice to make a customized one for our needs. In this chapter, we will utilize some of the common OpenSCAD libraries to design a laptop riser, which we will then print out.

In this chapter, we will cover the following:

- Designing the frame in Inkscape and OpenSCAD
- Designing the threaded rod in OpenSCAD
- Printing out our laptop stand

Technical requirements

The following are required to complete the chapter:

- Any late model Windows, macOS, or Linux computer that can install OpenSCAD and Inkscape.

- A 3D printer with PLA – any FDM printer should work, however, the Creality Ender 3 V2 is the printer used for our example.

- 4 X M3 10mm bolts

- A 3mm drill tap for making threaded holes, as shown here:

Figure 8.1 – 3mm drill tap

- M10 nylon cap nuts or another type of M10 nut, as shown here:

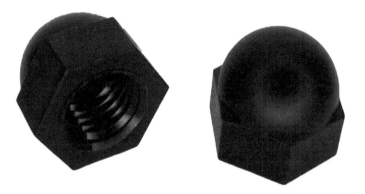

Figure 8.2 – M10 nylon cap nuts

The code and images for this chapter can be found here: `https://github.com/PacktPublishing/Simplifying-3D-Printing-with-OpenSCAD/tree/main/Chapter8`.

Designing the frame in Inkscape and OpenSCAD

Inkscape is a free and powerful design tool that we can use to create .svg files to import into OpenSCAD. In this section, we will download and install Inkscape. We will then take a very brief look at the tools we will be using. This will in no way be a comprehensive tutorial on Inkscape as it is a program that requires a fair bit of time to master.

We will use Inkscape to design the sides, or the frame, of our laptop stand. We will then import the file into OpenSCAD and modify it using code. We will export the file to an .stl file, so that we can 3D print it. We will start by downloading and installing Inkscape.

Downloading and installing Inkscape

The current version of Inkscape as of this writing is 1.1.2. It can be downloaded from the Inkscape website – http://inkscape.org. There are versions for Linux, Windows, and macOS. There is an option to download the source code as well.

To download and install Inkscape, we do the following:

1. Using a web browser, navigate to the Inkscape website at http://inkscape.org.
2. Click on **Download | Current Version**.
3. Select the appropriate version (**GNU/Linux, Windows, macOS**) by clicking on the corresponding button.
4. Follow the steps to download and install. For example, in Windows, there is an option to download a **64-bit** or **32-bit** version. Clicking on either one of these will give another option for **Installer in .exe format**, **Windows Installer Package**, or **Compressed archive in 7z format**. For our example, we will choose the **64-bit Windows Installer Package**.
5. Follow the usual steps to install the software.

Now that we have Inkscape installed on our computer, it is time to take a brief look at the tools we will be using to create the frame of our laptop stand.

Exploring Inkscape

When running Inkscape for the first time, a **Welcome!** screen will be presented. To set up Inkscape so that we can start to design, do the following:

1. Open Inkscape by clicking on its shortcut.
2. Click on **Save** to use Inkscape with the default settings.

3. A screen describing the Inkscape organization and how to donate may pop up. Click on **Thanks** once the information has been reviewed.

4. Click on **New Document** to open a blank document.

We should see a blank document. Before we start our design, we will set the canvas to the same size as the bed on the Ender 3 V2, so that we can view how our design will fit. To do this, we do the following:

1. From the top menu, click on **File | Document Properties…**.

2. Observe that the **Document Properties** dialog box pops up on the right side of the screen.

3. In the **Custom size** box, set **Width** to 220 mm and **Height** to 220 mm to correspond to the size of the Ender 3 V2 build plate, and press *Enter*:

Figure 8.3 – Setting a custom size in Inkscape

4. Observe that the size of the canvas has changed from a rectangle to a square. The canvas is now set to the same size as the Ender 3 V2 build plate.

Now that we have the canvas size set, let's start the design of the frame for our laptop stand.

Using Inkscape to design the frame

The frame consists of two equal but mirrored triangular-type shapes, which make up the sides of our laptop stand. They are held in place by threaded rods, which we will design in the upcoming *Designing the threaded rod in OpenSCAD* section.

We will use Inkscape to design the frame and expand our knowledge of Inkscape along the way. We will then import our Inkscape file into OpenSCAD where it will be used to create a 3D version of the frame. We will start with a basic shape.

Creating the basic shape

The design of the frame will start with a rectangle created with the rectangular tool in Inkscape:

1. Open up Inkscape with a new document.

2. Click *R* on the keyboard to draw a rectangle. Draw a rectangle of any size on the canvas (ignore rounded corners if present as we will be adjusting this property in *Step 3*):

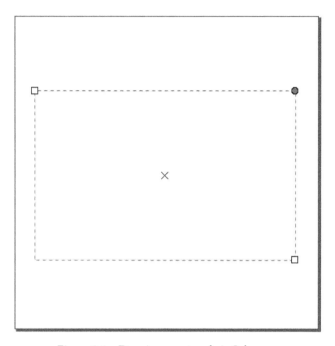

Figure 8.4 – Drawing a rectangle in Inkscape

3. Using the **Change** properties located at the top-left part of the screen, change the properties of the rectangle to the following:

Figure 8.5 – Change properties for the rectangle

W and **H** refer to the width and the height of the rectangle respectively. As we can see, our rectangle extends just outside of the canvas (the build plate size of the Ender 3 V2). This will not be a problem as the frame will be rotated to fit when we start to print out our design in the *Printing out our laptop stand* section. The **Rx** and **Ry** values refer to the roundness of the corners of our rectangle. Setting both values to 10 gives us the desired starting shape. Hit *S* on the keyboard to hide the **Change** properties and select the rectangle.

4. We will now cut the rectangle diagonally. Hit *Ctrl* + *D* on the keyboard to create a duplicate of the rectangle.

5. As we are now working with a duplicate rectangle, the **Change** properties do not show up in the top-left corner. Instead, we will resize the duplicate using the **Width of selection** and **Height of selection** properties in the top middle of the screen. Change these values to the following:

Figure 8.6 – Width and height settings for the duplicate rectangle

6. We will use the duplicate rectangle to cut our original rectangle diagonally. To do this, we need to rotate and put the duplicate rectangle in place. Click on the duplicate to get rotation arrows:

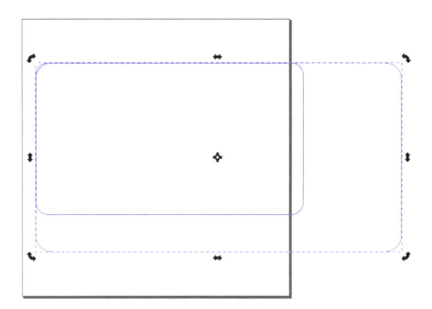

Figure 8.7 – Rotating the duplicate

7. Use the corner arrows to rotate the duplicate 45 degrees. Clicking on the duplicate again returns it to selection mode where it may be moved. Using these two techniques, position the duplicate such that it divides the original rectangle diagonally:

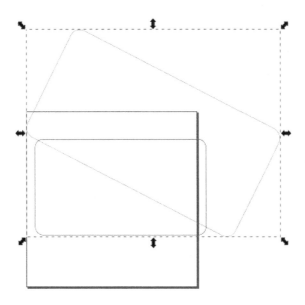

Figure 8.8 – Positioning the duplicate rectangle

8. Be sure to position the duplicate above the rounded corners of the original. Hit
    Ctrl + A on the keyboard to select both shapes. From the top menu, select **Path |
    Difference**. Observe that a new triangular shape is created:

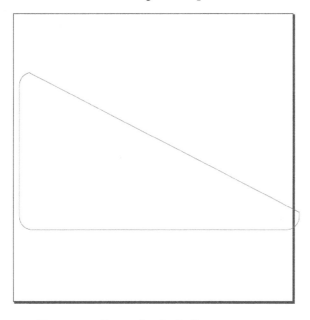

Figure 8.9 – Shape after the Difference operation

With the basic shape created, we will now modify it to make it more visually appealing.

Modifying the basic shape

We will modify the basic shape by curving the bottom and sides. To do this, we use the **Edit paths by nodes** tool:

1. Click *N* on the keyboard to select the **Edit paths by nodes** tool.

2. Select the shape by hitting *Ctrl + A* on the keyboard. Hover the mouse over the middle of the bottom line of the shape until it turns into a cursor with a four-sided arrow. Click on the left mouse button and drag up to curve the shape.

3. Do the same for the left side of the shape:

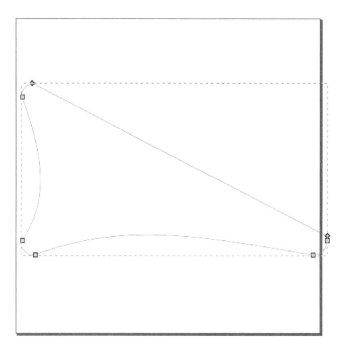

Figure 8.10 – Modifying the basic shape

4. Using the techniques from the steps in the *Creating the basic shape* section, draw a box on the bottom (*Figure 8.11 a*) and use it to flatten the bottom of our shape by using **Path | Difference** to make the bottom of the shape flatter (*Figure 8.11 b*):

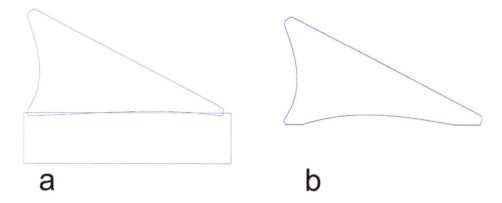

a   b

Figure 8.11 – Flattening the frame shape

5. Using the rectangular tool, create a box with a width of 10, a height of 20, and an **Rx** and **Ry** of 5. Place and rotate this box at the bottom front of the design, so that it is parallel to the flat part of our design:

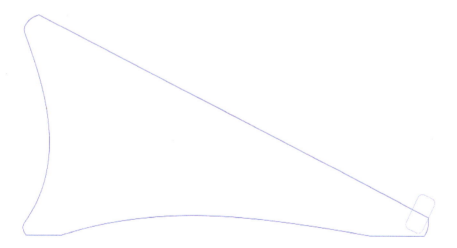

Figure 8.12 – Adding a box to the design

6. This box will serve as the edge that keeps our laptop from sliding. Hit *Ctrl + A* on the keyboard to select both shapes. From the top menu, select **Path | Union**. Observe that the two shapes have been combined into one shape to complete our design:

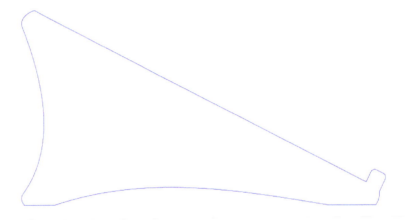

Figure 8.13 – Combining the shapes

7. With the Inkscape part of our design done, save the file as `frame.svg` to a directory where we will use it in OpenSCAD.

We will now import `frame.svg` into OpenSCAD to create a 3D version of the frame.

Using OpenSCAD to complete the design

In *Chapter 5*, *Using Advanced Operation of OpenSCAD*, we imported a `.svg` file into OpenSCAD to create a Thumbs Up! award. In *Chapter 6*, *Exploring Common OpenSCAD Libraries*, we looked at the `shell2D` operation from the Round Anything library. We explored arrays in *Chapter 7*, *Creating a 3D-Printed Name Badge*. We will use these three concepts to create the 3D version of the frame of our laptop stand. We will start with the `shell2D` operation:

1. Create a new file in OpenSCAD and save it to the same location as `frame.svg`.

2. Type the following code into the editor:

```
use <Round-Anything/roundAnythingExamples.scad>
$fn=200;
shell2d(0, -12)
{
    import("frame.svg",center=true);
}
```

3. After bringing in the `roundAnythingExamples.scad` library, we import our `frame.svg` file into a `shell2d` operation, and set it to create an inwards 12 mm shell. Save the file. Observe that the following shape has been created:

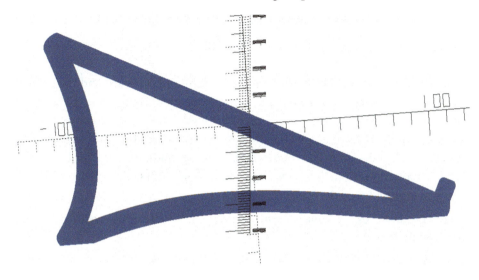

Figure 8.14 – Shell operation on the frame.svg import

4. To hold the two frame pieces together, we will require threaded rods. We will make a space for our threaded rods on the frame by adding circles. We will use arrays to store the values for the circles. This will allow us to quickly change values to get the circles in the right position. Change the code to the following:

```
use <Round-Anything/roundAnythingExamples.scad>
$fn=200;

thickness=5;
front_circle = [62,-42,12];
back_circle = [-80,-30,20];

linear_extrude(thickness)
union()
{
    shell2d(0, -12)
    {
        import("frame.svg",center=true);
    }
```

```
            translate([front_circle[0],front_circle[1]])
            circle(front_circle[2]);


            translate([back_circle[0],back_circle[1]])
            circle(back_circle[2]);
    }
```

We have added three variables to set the `thickness` of our part, and the position and size of two circles using arrays, `front_circle` and `back_circle`. The first value in the array sets the *x* position of the circle, the second value sets the *y* position, and the third value sets the diameter of the circle. The array values will be different for each design of the frame and should be modified if the circles do not sit in the correct position. The ones shown here work with the shape created in the *Using Inkscape to design the frame* section.

5. Save the file and click on **Render** or hit *F6* on the keyboard. Observe that the frame has been modified:

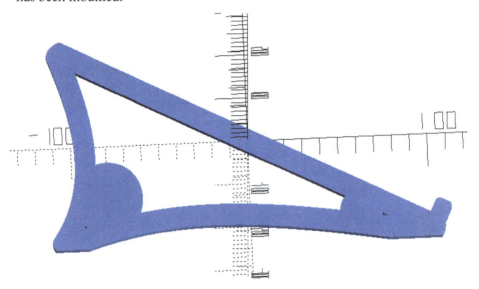

Figure 8.15 – Modified frame

6. With the circles in place, it is now time to add screw holes to the frame. The screw holes allow the threaded rods to connect the frame together. We will create a module for adding the screw holes. Add the following code after the variable declarations:

```
module create_screw_hole()
{
```

```
cylinder(d=11, h=500);

translate([0,0,-500])
cylinder(d=16, h=500);
}
```

This code creates an `11` mm hole with a `16` mm countersink. As we will strictly be using 10 mm hardware for our design, hardcoding the values here (instead of declaring them through variables) is acceptable. We give a `1` mm buffer to both the hole and countersink as our rod will be M10 (10 mm) with a `15` mm shoulder.

7. To complete our code, we will wrap up the remaining non-module code in a module itself. Replace the code after the `create_screw_hole()` module with the following:

```
module create_frame()
{
    difference()
    {
        linear_extrude(thickness)
        union()
        {
            shell2d(0, -12)
            {
                import("frame.svg",center=true);
            }
            translate([front_circle[0],
            front_circle[1]])
            circle(front_circle[2]);

            translate([back_circle[0],
            back_circle[1]])
            circle(back_circle[2]);
        }
        translate([front_circle[0],
        front_circle[1],
        thickness/2])
        create_screw_hole();
```

```
                translate([back_circle[0],
                back_circle[1],
                thickness/2])
                create_screw_hole();
        }
    }
create_frame();
```

What we are doing here is taking the difference between our frame and the screw holes. We can take advantage of the `front_circle` and `back_circle` array values to position the screw holes, as the holes are positioned in the center point of the circles.

8. Click on **Render** or hit *F6* on the keyboard. Observe that the screw holes and countersinks are added (the design may have to be rotated to see the countersinks):

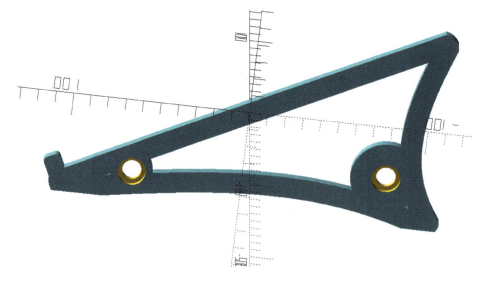

Figure 8.16 – Completed frame

9. With the design completed, hit *F7* on the keyboard to export the design as an `.stl` file. Save the file as `frame_left.stl` (as this will be the left side of the frame).

As we will be using the mirror functionality in Cura, there is no need to create an `.stl` for the right frame. It is now time to create the threaded rod that will hold our laptop stand together.

Designing the threaded rod in OpenSCAD

Creating 3D printed threads and nuts can be challenging due to the imperfections in the process. A slight droop while printing can result in a threaded connection that is too tight to be of use. Generally, this occurs with smaller-sized threads such as M3 or M4. For this reason, we will only use 3D print threads on the rods for our laptop stand and use standard M10 nuts to secure the rods to the frame.

Why not print all the hardware we need?

It is of little value to 3D print objects that are available in great abundance, such as standard nuts and bolts. The cost of the filament coupled with the time taken makes printing standard nuts and bolts expensive compared to just purchasing the hardware. This, of course, would not be the case for those who live in more remote areas.

For the threaded rod in our laptop stand design, we will be using a library to generate an M10 bolt. We will replace the head of the M10 bolt with a long cylinder and construct our threaded rod by connecting two rods with a connector plate. Designing with a connector plate makes the part easier to print as it removes the overhangs of a dual threaded rod. Also, the connector plate will act as a sort of raft during printing, as its surface area is greater than the rod itself. The rods will be joined together using M3 10 mm bolts. The threaded holes for these bolts will be tapped into the connector plates using an M3 drill tap.

We will start by using the `threads.scad` library to create an M10 bolt before we add a cylinder and connector plate to complete the design.

Creating a rod with an M10 threaded top

There is an operation in the `threads.scad` library called `MetricBolt`, which creates a metric bolt of a specified size. We will use this operation to create a thread to add to a cylinder for our rods.

To create the thread, we do the following:

1. Create a new file in OpenSCAD and type in the following:

```
use <threads.scad>
$fn=200;
translate([0,0,-10])
difference()
{
    MetricBolt(10, 10, tolerance=0.4);
```

```
        cylinder(h=10, d=100);
    }
```

What we are doing here is bringing in the `threads.scad` library to get the `MetricBolt` operation. We take the difference between an M10, 10 mm bolt and a `cylinder` that is 10 mm in height and `100` mm in diameter. The height of the `cylinder` corresponds to the height of the head of the `MetricBolt` (10 mm by default and not settable). The `difference` operation will leave us with just the thread. We then move the thread down 10 mm using a `translate` operation to place it at `0` on the *z* axis.

2. Click on **Render** or hit *F6* on the keyboard. Observe that only the thread is shown:

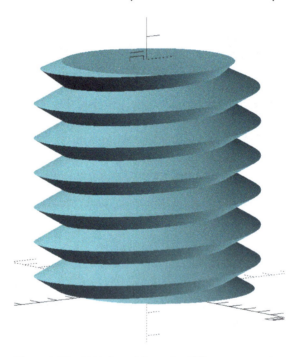

Figure 8.17 – M10 thread from the difference operation

3. We will now add a cylinder to the thread to serve as the shoulder of the rod. The cylinder will be `15` mm in diameter and will provide a snug fit with the countersink on the frame. We will put our code in a module to make it easier to handle. Change the thread code (the code starting with the translate operation) to the following:

```
module create_rod(height)
{
    translate([0,0,-10])
```

```
    difference()
    {
        MetricBolt(10, 10, tolerance=0.4);
        cylinder(h=10, d=100);
    }
    translate([0,0,-height/2])
    cylinder(h=height, d=15, center=true);
}
create_rod(150);
```

Our new `create_rod()` module adds a `cylinder` to the thread with a height determined by the `height` parameter. The `cylinder` has a set diameter of 15 mm, which makes it 5 mm larger than the thread. We call the `create_rod()` module with a value of 150 to create a cylinder with a height of 150 mm from the bottom of the thread.

4. Click on **Render** or hit *F6* on the keyboard. Observe that the rod generated consists of a thread and cylinder:

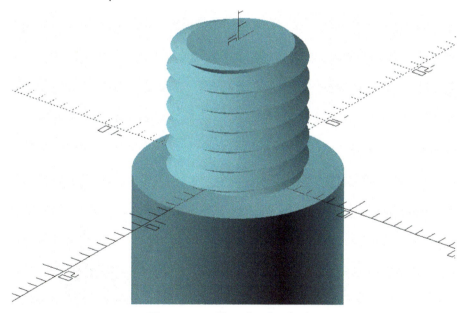

Figure 8.18 – Thread with cylinder

For our design, we will require four rods as we will be joining two together to form a longer rod. To do this, we require a connector plate. Let's now create the connector plate.

Adding a connector plate

We will build the connector plate with three modules – the first module to create the basic shape, the second module to create the screw holes, and the third module to bring the first two together to create the plate. We will then modify the create_rod() module to include code to add the connector plate to the rod.

We will start with a basic shape.

Creating the connector plate

To create the basic shape for our connector, we will use the intersection operation on a circle and square:

1. Comment out the line create_rod(150); from the code we added in *Step 3* of the *Creating a rod with an M10 threaded top* section.

2. Add the following module above the create_rod() module:

```
module create_connector_plate_shape()
{
    circle_radius=14.5;
    move_x = 20;
    width = 40;
    height = 20;

    intersection()
    {
        circle(circle_radius);
        translate([move_x,0])
        square([width, height], center=true);
    }
    mirror([1,0,0])intersection()
    {
        circle(circle_radius);
        translate([move_x,0])
        square([width, height], center=true);
    }
}
```

We are taking the `intersection` of a `circle` and a `square` based on parameters set at the top of the module. The values of `circle_radius`, `move_x`, `width`, and `height` are all hardcoded values, as our rod will always be based on an M10 thread and a `15` mm diameter rod. We use the `mirror` operation to create a symmetrical shape.

3.   Add the following line of code to the bottom:

```
create_connector_plate_shape();
```

4.   Click on **Render** or hit *F6* on the keyboard to observe the following shape:

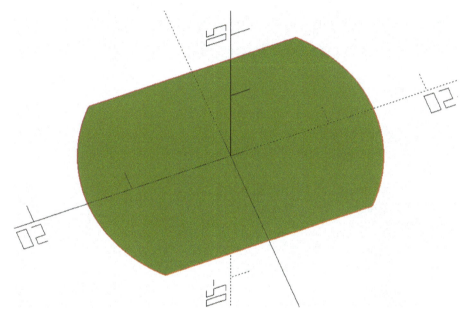

Figure 8.19 – Connector plate shape

5.   To create the screw holes to join two rod pieces together, we use another module. Add the following module below the `create_connector_plate_shape()` module:

```
module create_screw_hole()
{
    translate([0,0,-250])cylinder(d=2.5, h=500);
}
```

> **Is the order of the modules important?**
>
> As we have gone through the steps, there have been instructions to put modules in certain places. This is only to make the code more readable as the order of the modules is not important. However, the order of the code inside a module is important, as it will be run from top to bottom. Generally, it is a good practice to keep modules that make the final part such as `create_rod()` at the bottom and "support" type modules such as `create_connector_plate_shape()` at the top.

6. To create the connector plate, we simply take the `difference` from an extruded `create_connector_plate_shape()` and two calls to `create_screw_hole()`. Add the following module below `create_screw_hole()`:

```
module create_connector_plate()
{
    difference()
    {
        linear_extrude(4)
        create_connector_plate_shape();

        translate([11,0,0])
        create_screw_hole();

        translate([-11,0,0])
        create_screw_hole();
    }
}
```

7. Replace the line `create_connector_plate_shape();` from *Step 3* with the following:

```
create_connector_plate();
```

8. Click on **Render** or hit *F6* on the keyboard. Observe that a connector plate with screw holes is generated:

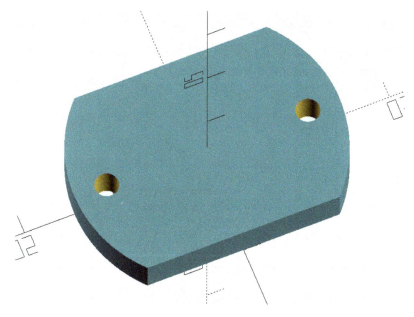

Figure 8.20 – Generated connector plate

With the connector plate code in place, it is now time to modify the create_rod()
module to complete the design of the rod.

Attaching the connector plate to the rod

To add the connector plate to the rod, we need to modify the create_rod() module.
To do this, we do the following:

1. Modify the code in the create_rod() module to the following:

```
module create_rod(height)
{
    translate([0,0,-10])
    difference()
    {
        MetricBolt(10, 10, tolerance=0.4);
        cylinder(h=10, d=100);
    }
    translate([0,0,-height/2])
    cylinder(h=height, d=15, center=true);

    translate([0,0,-height])
```

```
        create_connector_plate();
    }
```

All we are doing here is adding the lines to create a connector plate moving it down in the *z* direction so that it sits at the bottom of the rod.

2. Uncomment out the call to `create_rod()`; and comment out `create_connector_plate();`:

```
    create_rod(150);
```

3. Click on **Render** or hit *F6* on the keyboard. Observe that a rod with M10 threads at one end and a connector plate at the other is generated:

Figure 8.21 – Rod with threads and a connector plate

4. With the rod design completed, hit *F7* on the keyboard to export the design as a `.stl` file. Save the file as `rod.stl`.

We will need four rods for our design as we will connect two together using the connector plate with M3 10 mm bolts. To prepare the screw holes for the M3 bolts, we will use an M3 tap (see *Figure 8.1*) to create threads through the screw holes. We will use standard M10 nylon cap nuts (see *Figure 8.2*) to attach the rods to the frames.

Before we do all that, we will need to print out the frames and rods. We will do that next.

Printing out our laptop stand

We will require two frames, four rods, four M10 nylon cap nuts, and 4 M3 10 mm bolts to build our laptop stand. We will 3D print the frames and rods. The rods will be held together with the M3 10 mm bolts.

Slicing and printing the frame

We will start by printing the left side of the frame. We will then mirror the frame in Cura and print it again. Let's begin:

1. Open up Cura and select **Generic PLA, 0.4mm Nozzle**:

Figure 8.22 – Generic PLA template

2. If prompted with **Discard or Keep changes**, click on the **Discard changes** button.

3. Under **Print Settings | Build Plate Adhesion**, set **Build Plate Adhesion Type** to **None**.

4. Click on **File | Open File(s)...** and select the frame_left.stl file created in the *Designing the frame in Inkscape and OpenSCAD* section.

5.   Using the **Rotate** and **Move** tools, position the frame so that it fits onto the bed. The frame will need to be flipped over so that the countersinks can be seen:

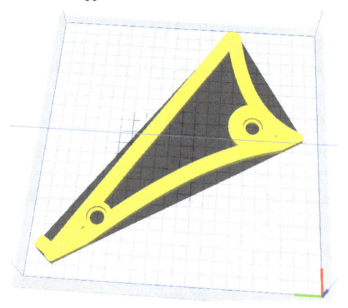

Figure 8.23 – Left side of the frame in Cura

6.   We will make three changes in **Print Settings**. Under **Top/Bottom**, set **Top/Bottom Thickness** to 1.2mm and click on **Enable Ironing**. Set **Infill Line Multiplier** under **Infill** to 2:

Figure 8.24 – Modifications to Print Settings

7. With these modifications, our frame should be quite sturdy (ironing not only adds a little extra filament, but it smooths out the top surface as well). Slice the file by clicking on the **Slice** button.

8. Insert a microSD card into the computer and click on the **Save to Removable Drive** button. Verify that the name of the file is `frame_left.gcode`.

9. Load the microSD card into the 3D printer and run the print job `frame_left`.

10. To create the right side of the frame, click on the frame in Cura and press the *M* key on the keyboard. Observe that the mirror arrows appear over the frame:

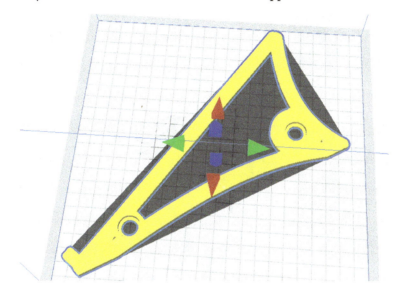

Figure 8.25 – Mirroring the frame

11. Click on a green arrow to mirror the frame.

12. Using the **Rotate** and **Move** tools, reposition the frame so that it fits on the build plate.

13. Change the name of the project to `frame_right` in the text area at the bottom-left of the screen:

Figure 8.26 – Changing the name to frame_right

14. Repeat *Steps 7 to 9* for the mirrored frame.

With the two frame pieces made, it is now time to 3D print the rods. We will print all four rods together.

Slicing and printing the rods

Adding the connector plates to the rods gives us a flat surface for the build plate. We will slice and print all four rods at the same time. We will clear the build plate and then add the rods. To do so, we do the following:

1. In Cura, select all objects on the build plate by hitting *Ctrl + A* on the keyboard.

2. Hit the *Delete* key on the keyboard to clear the build plate.

3. Click on **File | Open File(s)...** and select the file rod.stl.

4. Hit *Ctrl + A* and then *Ctrl + M* to get the **Multiply Selected Model** dialog:

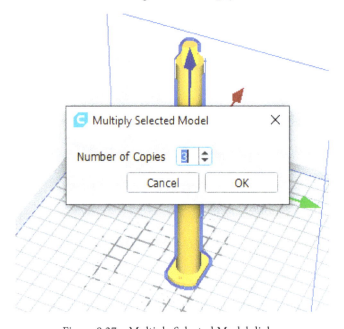

Figure 8.27 – Multiply Selected Model dialog

5. Enter the number **3** and click on the **OK** button.

6. Hit *Ctrl + A* and then *Ctrl + R* on the keyboard to auto arrange the models on the build plate.

7. We will change two **Print Settings** properties; turn off **Enable Ironing** and set **Build Plate Adhesion Type** to **Brim**:

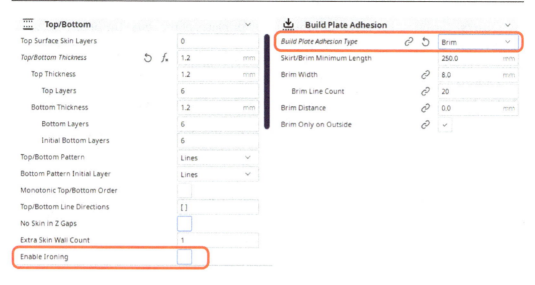

Figure 8.28 – Print Settings for rods

8. Slice the file by clicking on the **Slice** button.

9. Click on the **Preview** button to preview the print job:

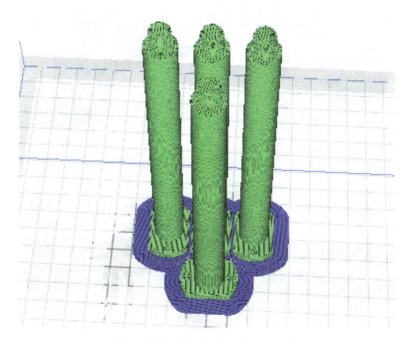

Figure 8.29 – Previewing the print job

10. Observe the brim that surrounds the connector plates of the rods. The brim increases the surface area that touches the build plate and provides strong adhesion during printing. Insert a microSD card into the computer and click on the **Save to Removable Drive** button. Verify that the name of the file is rod.gcode.

11. Load the microSD card into the 3D printer and run the print job rod.

With the frames and rods printed, it is now time to put the laptop stand together.

Putting the laptop stand together

To put the laptop stand together, we first add threads to the connector plates of the rods. We will then connect the rods together with M3 10 mm bolts. Using M10 nuts, we will attach the rods to the frames.

Let's start by taping out 3 mm threads on the connector plates:

1. Using a drill tap, as shown in *Figure 8.1*, tap a 3 mm thread into all eight of the connector plate screw holes:

Figure 8.30 – Adding a thread to the connector plate

2. Create two rods out of the four rod pieces using the M3 10mm bolts:

Figure 8.31 – Screwing the rods together

3. Insert the threaded end of the rod through the hole in the frame such that the countersink is on the inside:

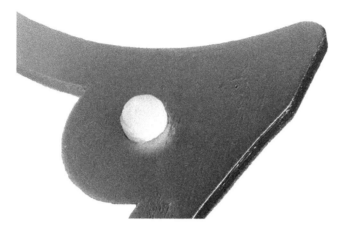

Figure 8.32 – Threaded rod and frame

4. Using the M10 nut, secure the rod to the frame. Do the same for all the holes in the frame:

Figure 8.33 – Completed laptop stand (painted)

This completes the construction of the laptop stand. With our design and the dimensions, our laptop stand should be useful for a wide range of 15" laptops. Adjustments can be made, however, for smaller laptops, such as Macbooks.

Summary

In this chapter, we created a laptop stand to use to elevate a laptop on a desk. We used the graphics program Inkscape to design the basic frame for our laptop stand. We then imported the Inkscape file into OpenSCAD, where we shelled it out using the `shell2D` operation before extruding it into a 3D shape.

We modified the frame by adding circles, which provided a place for screw holes for threaded rods. The threaded rods were designed in two pieces and were assembled with connector plates using M3 10 mm bolts. We made use of readily available M10 nuts to finish the construction of our laptop stand.

The main takeaway from this chapter is the use of 3D printing with traditional building techniques to make objects. Although we could've 3D printed threads for the connector plates and 3D printed M10 nuts, using a 3 mm tap and standard M10 nuts is far more effective.

In the next chapter, we will continue learning OpenSCAD as we design and 3D print a model rocket.

9
Designing and Printing a Model Rocket

On October 4th, 1957, the Soviet Union became the first country on Earth to launch a satellite into orbit with Sputnik 1. This set in motion a space race between the Soviet Union and the United States, which ultimately saw American men walk on the moon on July 20th, 1969. The hobby of model rocketry was born from this era. Rockets designed and built from lightweight materials such as plastic, balsa, and paper made model rocketry a safe and educational endeavor, inspiring many young people to opt for STEM (science, technology, engineering, and math) fields. In fact, many credit this time in history for the amazing technological innovations we have today.

Early model rockets were built using paper tubes, lathe-spun balsa nose cones, and hand-cut balsa fins. In this chapter, we will use 21st-century 3D design and 3D printing technology to create our own model rocket from a discarded paper towel tube.

We will cover the following topics:

- Creating the motor mount
- Creating the nose cone
- Creating the fins
- Assembling and launching the model rocket

Technical requirements

The following is required to complete the chapter:

- Any late-model Windows, macOS, or Linux computer that can install OpenSCAD and Cura.
- 3D printer with ABS and PLA – any FDM printer should work; however, the Creality Ender 3 V2 is the printer used for our example.
- Paper tube from paper towel roll (referred to as the body tube).
- 3 mm drill tap.
- 2 M3 10 mm bolts.
- Digital caliper for measurement.
- Epoxy glue and popsicle stick.
- 1 meter of elastic cord.
- Model rocket parachute – `https://bit.ly/3pVz1aY` or `https://www.youtube.com/watch?v=Y3xhZpmmboE`.

The code and images for this chapter may be found here: `https://github.com/PacktPublishing/Simplifying-3D-Printing-with-OpenSCAD/tree/main/Chapter9`.

Creating the motor mount

The motor mount of the model rocket is used to secure the model rocket motor in place inside the body tube. We will be constructing the motor mount from two rings connected by posts. The top ring will have a hole for the ejection charge of the rocket motor. The bottom ring will have a hole for the nozzle at the bottom of the model rocket motor. We will start our motor mount design by taking measurements of the paper tube.

Building around the paper tube

For those familiar with model rocketry, it is quite often the case that a simple paper tube from a paper towel roll does not fit with existing model rocket parts. However, with OpenSCAD and a 3D printer, we may custom build parts to fit any paper tube. We will design our nose cone, fin can, and motor mount around the paper tube, starting with the motor mount.

To design the motor mount, we need to get an accurate measurement of the internal diameter of the paper tube. To do this, we will build a measurement tool to assist us. We will start with a rough measurement of the internal diameter using a digital caliper.

Measuring the paper tube

Due to the lack of rigidity of paper tubes, getting exact measurements is a difficult thing to do. Any measurement we do get will be affected by how much we distort the paper tube while taking the measurement. The number we get will not be exact; however, it will be close enough to use as a starting point in our design.

To get the first measurement, use a digital caliper to measure the internal diameter of the paper tube. Be sure not to squeeze the paper tube to prevent distorting, as shown in the following figure:

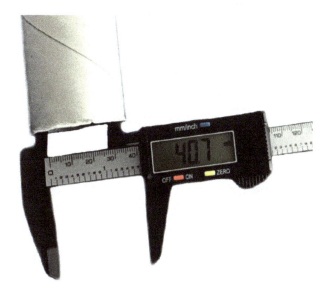

Figure 9.1 – Measuring the internal diameter of a paper tube

236 Designing and Printing a Model Rocket

For our tube, we measured an internal diameter of 40.7 mm. We will use this value to write code to generate the measurement tool.

Designing the measurement tool

The tool we will build to measure the internal diameter of the paper tube will be a cone-shaped cylinder. By inserting this shape into the paper tube, we may get a more accurate measurement than with our previous method. To build the measurement tool, we do the following:

1. Open up OpenSCAD and create a new file.

2. For our example, the internal diameter was measured at 40.7 mm. We will use this value in our code. Add the following code to our new OpenSCAD file:

```
$fn=200;
d_tool=40.7;
d_tool_offset=3;
tool_height=60;
module create_measurement_tool()
{
     cylinder(h=tool_height,
     d1=d_tool+d_tool_offset,
     d2=d_tool-d_tool_offset);
}
```

In our code, we have a module called create_measurement_tool() to create a cone-shaped cylinder with one end 3 mm larger than the measured value (d_tool) and the other 3 mm smaller than the measured value. We set the height of our tool to 60 mm using the tool_height variable. To generate the measurement tool, add the following code in the editor:

```
create_measurement_tool();
```

3. Click on **Render** or hit *F6* on the keyboard to observe the following:

Figure 9.2 – Tool to verify paper tube internal diameter

4. Hit *F7* on the keyboard and save the `.stl` file to the computer. Give it a descriptive name such as `measurement_tool.stl`.

With the `.stl` file saved to our computer, it is now time to print out and use the measurement tool to get an accurate internal diameter of the paper tube. We will use Cura to slice our object to G Code.

Printing out and using the measurement tool

We will print out our measurement tools in a light-colored **Polylactic Acid** (**PLA**). We will use a raft to avoid the elephant's foot effect and to provide strong bed adhesion.

What Is the Elephant's Foot Effect in 3D Printing?

The elephant's foot effect in 3D printing refers to the warping of the first few layers during a print job due to a lack of space between the print bed and nozzle. This warping takes its name from the shape of an elephant's foot. Often it is necessary to have the first layer "pushed" onto the print bed to gain adhesion, which can lead to the elephant's foot effect.

We will start by modifying a generic PLA profile:

1. Open up Cura and select the generic PLA profile by selecting **Material | Generic | PLA**:

Figure 9.3 – Generic PLA material profile

2. Set **Build Plate Adhesion Type** to **Raft**.

3. Using what we learned in *Chapter 3, Printing Our First Object*, print out the measurement tool using a light-colored PLA.

4. Once printed, insert the measurement tool into the paper tube and make a mark on the measurement tool right where it meets the paper tube (see *Figure 9.4 a*).

5. Using the digital caliper, take a measure where the marking is made. This will be the value we use to create our motor mount (see *Figure 9.4 b*):

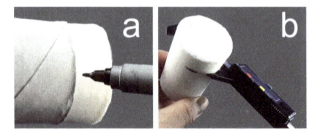

Figure 9.4 – Taking measurements of the paper tube using the measurement tool

With an accurate measurement of the internal diameter of the paper tube, we may now design and print the motor mount.

Designing and printing the motor mount

By using a standard paper tube from a paper towel roll, we will be making a model rocket that is considered larger than most model rockets. To successfully launch this rocket, we will require a C- or D-size model rocket motor, and thus our motor mount must accommodate for this.

> **What Does the Letter of a Model Rocket Motor Mean?**
>
> The letter used to classify a model rocket motor refers to the motor's total impulse range. Total impulse is the maximum momentum for a motor measured in newton-seconds. One newton-second is used to describe a force of one newton on an object for one second. A C motor has a total impulse range of 5.01–10 newton-seconds, while a D motor has a total impulse range of 20.01–40 newton-seconds. A good source of information for model rocketry is the Sigma Rockets YouTube channel (which yours truly helped to create): `http://www.youtube.com/sigmarockets`.

As C motors are generally 18 mm in diameter and D motors 24 mm in diameter, we will design our code to be dynamic enough to generate a motor mount for either size. We will use a conditional statement in our code to do this.

Writing code to generate a motor mount

With an accurate internal diameter measurement, we may now proceed to design the motor mount. We will design the motor mount to fit a standard 18 mm model rocket motor by default. We will use an `if` statement in our code to modify the motor mount to suit a 24 mm model rocket motor if desired.

We will parameterize our code with the following values:

- `d_actual` – The internal diameter of the paper tube as measured using the measurement tool.

- `motor_height` – The height of the model rocket motor. This value is 70 mm for both 18 mm and 24 mm motor rocket motors.

- `motor_diameter` – The default model rocket motor diameter. We will add an additional 0.5 mm to this value.

- `thrust_ring` – The thrust ring is used to keep the model rocket motor from moving up the paper tube. Its size, coupled with the `motor_diameter`, matches the thickness of the tube used in the model rocket motor. We will use a default value of 14 mm for an 18 mm model rocket motor.

- `ring_thickness` – This is the thickness of the top and bottom rings used in the motor mount.

- `screw_hole_distance` – This is the distance of the screw hole from the center. M3 10 mm bolts will be used to secure the model rocket motor in place.

- `mount_post_diameter` – This is the diameter of the posts that will connect the top and bottom rings together.

The following diagram illustrates the parameters used in designing our motor mount:

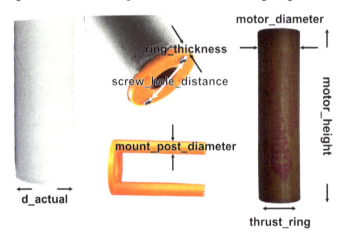

Figure 9.5 – Motor mount dimensions (not to scale)

Now we will start writing the code to create the centering ring, followed by motor mount posts.

Creating the centering ring

The centering ring is the centering adapter between the model rocket motor and the paper tube. For our motor mount, we will require two centering rings. In our code, we will define the variables used for the entire motor mount before writing the code for the centering ring.

We will start by creating a new OpenSCAD file:

1. Open up OpenSCAD and create a new file.

2. At the top of the file, add the variable declarations:

```
$fn=200;
d_actual=41.7;
motor_height=70;
```

```
motor_diameter=18.5;
thrust_ring=14;
ring_thickness=4;
screw_hole_distance=16;
mount_post_diameter=15;
```

3. In the editor, add the following code:

```
module create_centering_ring()
{
    difference()
    {
        cylinder(h=ring_thickness, d=d_actual);

        translate([0,0,ring_thickness/2])
        cylinder(h=500, d=motor_diameter);
        cylinder(h=500, d= thrust_ring, center=true);
    }
}
create_centering_ring();
```

4. With this code, we create a ring with an internal indent equal to the diameter of an 18 mm model rocket motor (plus 0.5 mm). Click on **Render** or hit *F6* on the keyboard and observe the centering ring:

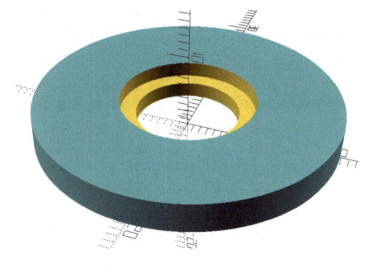

Figure 9.6 – 18 mm centering ring

5. As we want our code to produce 24 mm model rocket motors as well, we will add a new variable called `type`. If `type` is equal to `18` (as in 18 mm), we will generate an 18 mm centering ring (as in *Figure 9.6*); if not, we will generate a 24 mm centering ring. Change the code for the module `create_centering_ring()` and the call to this module to the following:

```
module create_centering_ring(type=18)
{
    difference()
    {
        cylinder(h=ring_thickness,d=d_actual);
        if(type==18)
        {
            translate([0,0,ring_thickness/2])
            cylinder(h=500,d=motor_diameter);

            cylinder(h=500,d=thrust_ring,
            center=true);
        }
        else{
            translate([0,0,ring_thickness/2])
            cylinder(h=500,d=motor_diameter+6);

            cylinder(h=500,d=thrust_ring+6,
            center=true);
        }
    }
}
create_centering_ring(24);
```

6. We add 6 to the value of `motor_diameter` and `thrust_ring` if `type` is not equal to `18`. This will generate a centering ring for a 24 mm model rocket motor. Click on **Render** or hit *F6* on the keyboard and observe that a centering ring for a 24 mm model rocket motor is generated:

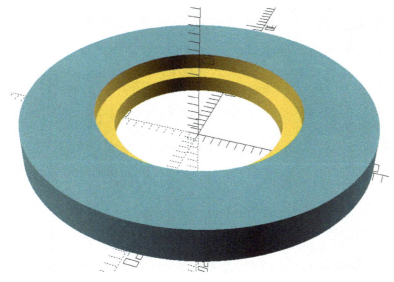

Figure 9.7 – 24 mm centering ring

With the code to generate the centering ring written, it is time to add the side posts that will hold the model rocket motor in the paper tube.

Creating the motor mount posts

We will add two posts to the centering ring. The posts will also be used to connect a bottom centering ring to the motor mount.

To write the code to create the motor mount post, do the following:

1. Delete the line `create_centering_ring(24);`.

2. Add the following module:

```
module create_mount_post(type=18)
{
    difference()
    {
        translate([screw_hole_distance,0,0])
        cylinder(d=mount_post_diameter,
        h=motor_height);
```

```
        if(type==18)
        {
            cylinder(d=motor_diameter, h=500,
            center=true);
        }
        else
        {
            cylinder(d=motor_diameter+6, h=500,
            center=true);
        }
        difference()
        {
            cylinder(d=1000, h=500,
            center=true);

            cylinder(d=d_actual, h=500,
            center=true);
        }
    }
}
```

With this module, we create a cylinder of diameter `mount_post_diameter` and a height of `motor_height`. We then move the cylinder in the *x* axis by the value of `screw_hole_distance` (we will use this variable to place the screw holes later in the code). We cut the post on one side with a cylinder equal to either `18` mm or `18 + 6` mm based on the value of `type`. We then use a hollow cylinder with an internal diameter equal to the internal diameter of our paper tube to create the final shape.

3. Add the following line to the code:

```
create_mount_post(24);
```

This will generate a motor mount post for a 24 mm model rocket motor.

4. Click on **Render** or hit *F6* on the keyboard and observe the following:

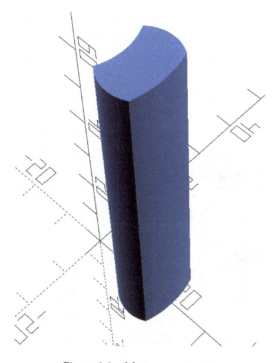

Figure 9.8 – Motor mount post

5. Before we add the post to the centering ring, we need a module for screw holes. Add the following module to the code:

```
module create_screw_holes(diameter)
{
    translate([screw_hole_distance,0,0])
    cylinder(d=diameter, h=500, center=true);

    translate([-screw_hole_distance,0,0])
    cylinder(d=diameter, h=500, center=true);
}
```

This module is very similar to the screw hole module from the *Adding a connector plate* section of *Chapter 8, Designing and Printing a Laptop Stand*. Basically, it creates a cylinder with a large height with a diameter determined by the diameter variable. It is moved in the *x* direction by the value of screw_hole_distance. The cylinder is centered to allow for a clean cut. It is then mirrored so that two screw holes may be generated.

6. We may now write the code to add a centering ring to motor mount posts. Add the following module to the code:

```
module create_main_bracket(type=18)
{
    difference()
    {
        union()
        {
            mirror([1,0,0])
            create_mount_post(type);

            create_mount_post(type);
        }
        create_screw_holes(2.5);
    }
    create_centering_ring(type);
}
```

The `create_main_bracket()` module generates two motor mount posts and adds 2.5 mm screw holes through a `difference()` operation. A centering ring is then added.

7. Replace the `create_mount_post(24);` line with the following:

```
create_main_bracket(24);
```

With this line, we will create a motor mount main bracket for a 24 mm model rocket motor.

8. Click on **Render** or hit *F6* on the keyboard and observe the following:

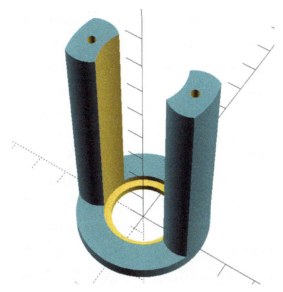

Figure 9.9 – Motor mount main bracket

9. Before we write the code to finish the motor mount, we will save the motor mount main bracket as an `.stl` file. Hit *F7* on the keyboard and save the file to the computer. Give it a descriptive name such as `main-bracket.stl`.

10. To complete the motor mount, we require a bottom centering ring to hold the motor in place. Add the following module to the code:

```
module create_bottom_bracket(type=18)
{
    difference()
    {
        create_centering_ring(type);
        create_screw_holes(3);

        rotate([0,0,90])
        create_screw_holes(6);
    }
}
```

The `create_bottom_bracket()` module generates a centering ring with two sets of screw holes. One set (3 mm) to screw the bottom bracket onto the main bracket and the other (6 mm) to act as air holes to help cool the model rocket motor after flight.

11. Replace the `create_main_bracket(24);` line with the following:

```
create_bottom_bracket(24);
```

With this line, we will create a bottom bracket for a 24 mm model rocket motor.

12. Click on **Render** or hit *F6* on the keyboard and observe the following:

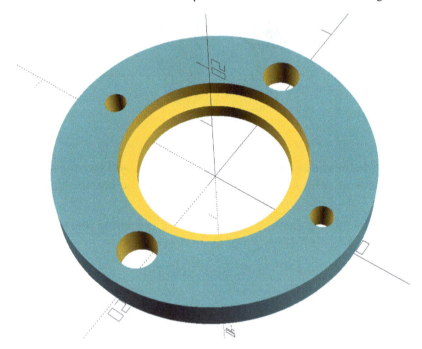

Figure 9.10 – Motor mount bottom bracket

13. Hit *F7* on the keyboard and save the file to the computer. Give it a descriptive name, such as `bottom-bracket.stl`.

With the bottom bracket generated and saved, we are now ready to print out the motor mount.

Printing out and installing the motor mount

As our motor mount will be touching the model rocket motor, we want it to be as resistant as possible to high temperatures. Ideally, we would print it with a liquid resin printer and an engineering-grade resin; however, as we are dealing with FDM printers, we will use a filament.

ABS melts at a higher temperature than PLA, is relatively inexpensive, and is readily available. Saying that, however, ABS can be a challenge to print with. An enclosure of some sort is desired to avoid the cracking that can occur as the ABS cools while printing (see *Figure 9.11*):

Figure 9.11 – ABS cracking

For our motor mount, small cracks would be acceptable as the motor mount will be inserted inside the paper tube. We will print with a draft shield to protect the print during the print job. An enclosure such as a tent enclosure is strongly recommended.

We will start by modifying a generic ABS profile:

1. Open up Cura and select the generic PLA profile by selecting **Material | Generic | ABS**.

2. Ensure that **All** settings are selected:

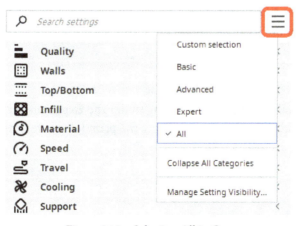

Figure 9.12 – Selecting All in Cura

3. Change the following default settings to the values shown here (as indicated by the red boxes):

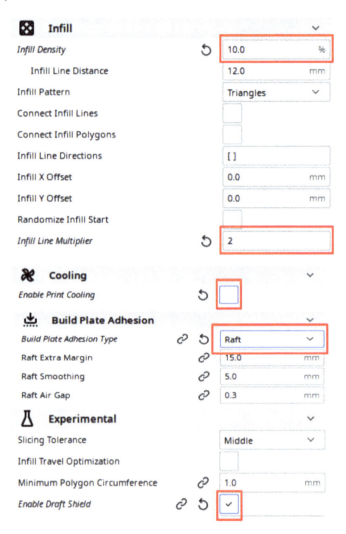

Figure 9.13 – Cura settings for ABS

Of note is the **Enable Draft Shield** setting under the **Experimental** section. This will create a shield around the print during the print job, protecting the print from cool air that might cause cracking.

4. Import the motor mount main bracket and motor mount bottom bracket into Cura using the **File | Open File(s)...** menu option.

5. For our example, we will load the 24 mm version of the motor mount parts. Click on the blue **Slice** button on the bottom-right part of the screen.

6. Click on the **Preview** button to observe what our print job will look like:

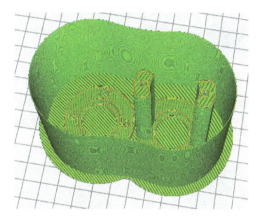

Figure 9.14 – Preview of the motor mount print job

7. Using what we learned in *Chapter 3, Printing Our First Object*, run the print job using ABS filament.

8. After the print is done, we will tap thread holes into the mount posts. Using a 3 mm drill tap, add threads to both posts:

Figure 9.15 – Adding threads to the motor mount

9. We will now glue the motor mount into the paper tube. Using a popsicle stick, spread a thin layer of epoxy glue about 60 mm inside the paper tube:

Figure 9.16 – Adding glue to the paper tube

10. Install the motor mount main bracket into the paper tube such that the motor mount posts are flush with the bottom of the paper tube:

Figure 9.17 – Installing the motor mount main bracket

11. Using M3 10 mm bolts, attach the motor mount bottom bracket to the motor mount main bracket. The bottom bracket will be removed and re-attached when installing the model rocket motor:

Figure 9.18 – Attaching the bottom bracket

With the motor mount installed, it is now time to take an accurate outside diameter measurement of the paper tube.

Getting an accurate outside diameter measurement

With the motor mount installed, an accurate measurement of the outside diameter of the paper tube may be taken by following these steps:

1. Using a digital caliper, take the outside diameter measurement against the inside centering ring:

Figure 9.19 – Taking the outside diameter measurement of the paper tube

2. In our example, we measure the outside diameter at 42 mm. We will add 0.5 mm to this value for a buffer.

We will use this measurement when we write the code to generate the nose cone and fins.

Creating the nose cone

The nose cone sits on top of the rocket and has a shape that reduces the aerodynamic drag on the rocket as it goes up. It attaches to the rocket through a shock cord and is jettisoned from the rocket when the ejection charge of the rocket motor is fired. A parachute attached to the shock cord brings the rocket down safely.

We will build the nose cone in two parts – the shoulder and the cone. Then we will print the nose cone. Let's start by building the shoulder.

Designing the nose cone

In the *Printing out and using the measurement tool* section, we measured an accurate internal diameter of the paper tube. We will use this value to create the shoulder part of the nose cone. We will also use the measured outside diameter taken in the *Getting an accurate outside diameter measurement* section.

The variables we will use to create the nose cone are the following:

- `diameter_in` – This is the internal diameter of the paper tube and will be used to create a nose cone shoulder.

- `diameter_out` – This is the outside diameter of the paper tube. It will be used for the cone shape part of the nose cone.

- `cone_height` – The height of the cone part of the nose cone or the part that sits outside of the paper tube.

- `shoulder_height` – The height of the shoulder part of the nose cone.

- `taper` – The value subtracted from `diameter_in` to create a taper at the bottom of the nose cone shoulder. The taper allows the nose cone to slide into the paper tube with ease.

We will start by creating a new file in OpenSCAD:

1. Open up OpenSCAD and create a new file.

2. At the top of the file, add the variable declarations:

```
$fn=200;
diameter_in=41.7;
diameter_out=42.5;
cone_height=65;
shoulder_height=20;
taper=1;
```

3. Add the following code after the variable declarations:

```
module create_shoulder()
{
    difference()
    {
        cylinder(h=shoulder_height,
        d1=diameter_in-taper,
        d2=diameter_in);

        rotate([0,90,0])
        linear_extrude(height=10, center=true)
        difference()
        {
```

```
                circle(15);
                circle(5);
            }
        }
    }
```

The `create_shoulder()` module creates a cylinder with a slight cone shape. The bottom diameter is set to 1 mm less (`taper`) than the internal diameter of the paper tube, and the top is equal to the internal diameter. A sideways ring is subtracted from this shape to provide a channel to thread the shock cord through.

4. Add the following code below the module:

```
create_shoulder();
```

5. Click on **Render** or hit *F6* on the keyboard to observe the nose cone shoulder:

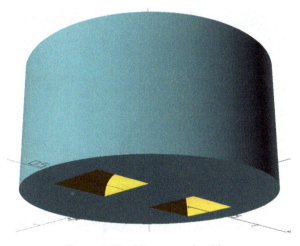

Figure 9.20 – Nose cone shoulder

6. With the nose cone shoulder in place, it is time to add the cone to the top. Add the following module to the code before the `create_shoulder();` line:

```
module create_cone()
{
    translate ([0,0,shoulder_height])
    resize([diameter_out,
    diameter_out,
    cone_height])
```

```
    difference()
    {
        sphere(d=diameter_out);
        translate([0,0,-500])
        cube([1000,1000,1000],center=true);
    }
}
```

The create_cone() module creates a cone by stretching a sphere by the value of cone_height. The diameter of the cone is set to the measure outside diameter of the paper tube (diameter_out). A large cube is used to remove the bottom part of the stretched sphere to create a cone. The entire shape is moved up by the value of shoulder_height to place it on top of the nose cone shoulder.

7. Add a call to create_cone() to the code. Our nose cone is created with the following two lines of code:

```
create_shoulder();
create_cone();
```

8. Click on **Render** or hit *F6* on the keyboard to observe the creation of the nose cone:

Figure 9.21 – Nose cone

9. Hit *F7* on the keyboard and save the file to the computer. Give it a descriptive name, such as nosecone.stl.

With the nose cone created, it is time to 3D print it.

Printing out the nose cone

With the nose cone `.stl` file created, we will now print out the nose cone. We will use PLA as it is easier to print with than ABS.

As in our previous print jobs, we will start with a generic template and modify it to our needs. We will use the generic PLA template:

1. Open up Cura and select the generic PLA profile by selecting **Material | Generic | PLA**.

2. Set **Infill Density** to `10%` and **Infill Line Multiplier** to 2 under the **Infill** settings:

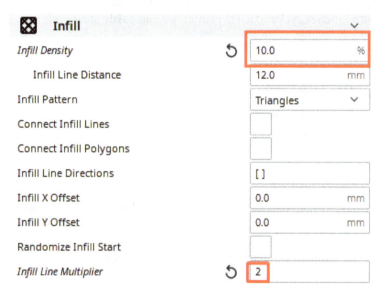

Figure 9.22 – Modified Infill settings

3. Load the nose cone into Cura using the **File | Open File(s)...** menu.

4. Click on the blue **Slice** button on the bottom-right part of the screen to create the print job.

5. Using what we learned in *Chapter 3, Printing Our First Object*, run the print job using PLA filament.

With the nose cone designed and printed, it is now time to create the fins for our model rocket. Let's do that now.

Creating the fins

Traditionally, fins for model rockets were cut from thin balsa sheets and glued to the body tube of the rocket. Early kits provided cut-out patterns from paper that were traced onto the balsa sheet and cut out with a sharp hobby knife. When laser cutters became available, these fins were pre-cut out of balsa sheets, making it faster and easier for the kit builder to build their model rocket.

Still, the challenge with balsa fins was in gluing them to the body tube as they took a long time to dry. Getting the fins straight also proved to be difficult at times.

Plastic fin cans (fins pre-attached to a tube) make putting fins on a model rocket much easier. Armed with OpenSCAD and a 3D printer, we can easily create our own fin cans.

Let's do that now.

Designing the fin can

Our fin can will be designed such that we may alter the number of fins generated. A "launch lug," or small tube to hold the rocket on the launchpad for the first meter or so of flight, will be built into the fin can.

What Is a Launch Lug?

A launch lug is a fancy name for a paper-coated straw that generally is glued midway onto the body tube of a model rocket. The paper coating allows it to be glued onto the body tube with standard white glue. For larger rockets, it is common to break the launch lug up into multiple pieces on the body tube. For our example, we will build the launch lug onto the fin can, to keep things simple.

For our design, we will be using the OpenSCAD **polygon** operation to define the fin shape. This operation allows us to create a shape by passing in *x* and *y* coordinates. We will also use a `for` loop to dynamically generate the number of fins we want.

The variables we will use are the following:

- `height` – The height of the cylinder of the fin can as well as the length of the leading edge of a fin

- `diameter` – The internal diameter of the fin can

- `thickness` – The value to determine the thickness of the fins, fin can cylinder, and launch lug

- `guide_diameter` – The diameter of the launch lug or guide
- `fins` – The number of fins that will be generated for our fin can
- `fin_shape` – The *x*, *y* coordinates to determine the shape of the fin

We will start by verifying the fin shape.

Generating the fin design using the polygon operation

The `fin_shape` parameter defines the coordinates to draw out the fin design. By using the `height` parameter inside the `fin_shape` coordinates, we ensure that the leading edge of the fin or the edge that connects to the fin can cylinder is always the same height as the tube.

Let's start the design of our fin can by viewing the shape of the fin. We will start with a new OpenSCAD file:

1. Open up OpenSCAD and create a new file.
2. At the top of the file, add the variable declarations:

```
$fn=200;
height=70;
diameter=43.5;
thickness=1.2;
guide_diameter=3;
fins=3;
fin_shape=[[0,0],[height,0],
          [100,35],[100,45],
          [35,35]];
```

3. With the variable declarations in place, let's verify the shape of the fin. We will use OpenSCAD's `polygon` operation on the `fin_shape` variable to do so. Type in the following code below the variable declarations:

```
polygon(fin_shape);
```

4.   Click on **Render** or hit *F6* on the keyboard to observe the shape of the fin:

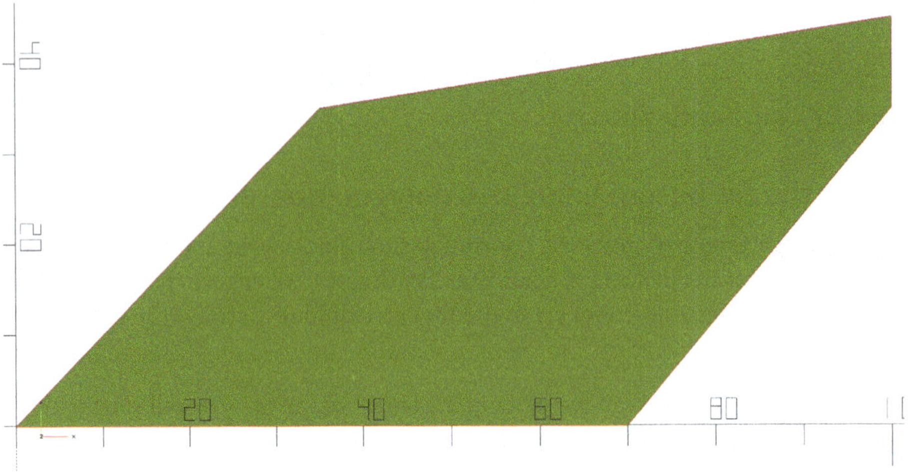

Figure 9.23 – Fin shape generated with the polygon operation

With our fin shape defined, let's create a module that will create a 3D version of the fin.

Extruding the fin design into a 3D shape

With the polygon operation, we can see a 2D shape of the fin design. To turn this design into a useful 3D shape, we simply extrude it and put the extruded shape in place with `rotate` and `translate` operations.

Let's do that with a new module in our code:

1.   Delete the `polygon(fin_shape);` line and replace it with the following module:

```
module create_fin(angle=0){
    rotate([0,0,angle])
    translate([diameter/2,0,height])
    rotate([90,90,0])
    linear_extrude(height=thickness, center=true)
    polygon(fin_shape);
}
```

To understand this code, let's work from the bottom to the top. The `linear_extrude()` operation makes a 3D version of the polygon of `fin_shape`. It is extruded by the value of `thickness` and is centered. The fin is rotated with the `rotate()` operation such that it sits upright in the *x* and *y* axis (`rotate([90,90,0])`). It is then moved in the *x* axis equal to the radius of the fin can and up in the *z* axis by the value of `height` (`translate([diameter/2, 0, height])`). The `rotate([0,0,angle])` operation rotates the extruded fin by an angle equal to the `angle` parameter (with a default value equal to 0) passed into the module in the *z* direction.

2. To test out this module, add the following line to the bottom of the code:

```
create_fin();
```

3. Click on **Render** or hit *F6* on the keyboard to observe the extruded fin. Observe how it has been rotated and then moved away from the *x* and *z* axes:

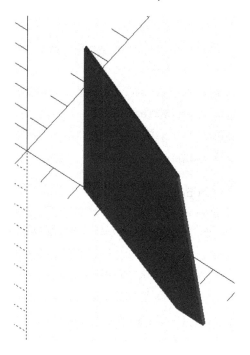

Figure 9.24 – Extruded fin

With the code to extrude and move the fin into place, it is now time to finish the design of the fin can. We will create a new module to do that.

Creating the fin can from extruded fins

We will now create a new module to generate the fin can using the `create_fin()` module. The `create_fin()` module will be called for every fin we need to generate as determined by the `fins` parameter (set to 3 in our variable declarations). To do so, we do the following:

1. Delete the `create_fin();` line and replace it with the following module:

```
module generate_fin_can()
{
    difference(){
        union()
        {
            cylinder(h=height,
            d=diameter+thickness);

            for ( i = [0 : fins-1] ){
                create_fin(i*(360/fins));
            }
            rotate([0,0,180/fins])
            translate([diameter/2+
                    guide_diameter/2+thickness/2,
                    0, 0])

            difference(){
                cylinder(h=height,
                d=guide_diameter+thickness);

                cylinder(h=height,
                d=guide_diameter);
            }
        }
        cylinder(h=1000, d=diameter, center=true);
    }
}
```

At the heart of this module is a `for` loop that places fins evenly around the fin can with the line `create_fin(i*(360/fins));`. For a three-fin rocket, this would be done every 120 degrees. A cylinder with a diameter equal to the outside diameter of the paper tube (`diameter`) plus `thickness` is added to the fins (the fin can cylinder). The second `difference()` operation creates the launch lug, which is added to the side of the cylinder (`translate([diameter/2+guide_diameter/2+thickness/2,0,0])`), moving it by the radius and half the thickness. A cylinder with a diameter equal to the outside diameter of the paper tube is used to cut a hole through the fins and cylinder assembly, thereby creating a fin can that can slide over the paper tube.

2. Type in the following code below the `generate_fin_can()` module:

```
generate_fin_can();
```

3. Click on **Render** or hit *F6* on the keyboard to observe the generated fin can:

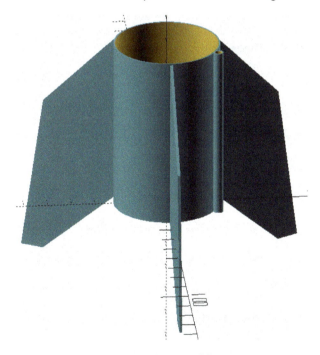

Figure 9.25 – Generated fin can

4. Hit *F7* on the keyboard and save the file to your computer. Give it a descriptive name, such as `fincan.stl`.

We will now 3D print the fin can, so that we can start assembling the rocket.

Printing out the fin can

Printing out the fin can may be challenging as most of its shape would sit above the build plate, thus requiring a lot of support material. Removing support material may cause damage to the fin can as the fins themselves are thin. Arguably the best way to print out the fin can is upside down with a raft to keep it from moving while printing. The raft should be removed without leaving any plastic behind. However, in the case where there is a little bit of the raft on the fin can, these bits can be sanded off.

We will use PLA as it is relatively easy to print with. We are not too concerned about the melting temperature as the fin can will be far enough from the rocket motor, therefore heat is not a concern. We will use Cura and a generic PLA template:

1. Open up Cura and select the generic PLA profile by selecting **Material | Generic | PLA**.

2. Modify the **Build Plate Adhesion** settings by setting **Build Plate Adhesion Type** to `Raft` and **Raft Extra Margin** to `20.0` mm and ensuring that **Raft Air Gap** is set to `0.3` mm:

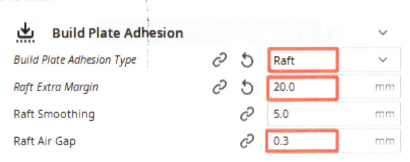

Figure 9.26 – Cura PLA modifications for fin can

What Is Raft Air Gap?

Raft Air Gap sets the gap between the last layer of the raft and the first layer of the fin can. Having this value at 0.3 mm creates a gap of 0.3 mm between the raft and the fin can. This should be sufficient to allow for easy removal of the raft. Raft Air Gap may be increased if the raft is difficult to remove or decreased if the fin can fails to stay on the raft during printing.

---

3. Load the fin can into Cura using the **File | Open File(s)...** menu.

4. Rotate the fin can 180 degrees vertically such that the fins are pointing upward:

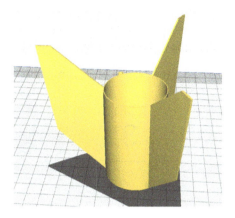

Figure 9.27 – Rotated fin can in Cura

5. Click on the blue **Slice** button in the bottom-right part of the screen to create the print job.

6. Using what we learned in *Chapter 3, Printing Our First Object*, run the print job using PLA filament.

With our fin can printed, we have all the 3D-printed parts we need to construct the rocket. We will now install a shock cord, parachute, rocket motor, and model rocket wadding before launching the rocket.

Assembling and launching the model rocket

Those familiar with model rocketry understand how a model rocket kit is constructed. Generally, the motor mount is installed first followed by the shock cord and fins. The nose cone and parachute are installed using the shock cord once the glue on the shock cord is dry.

The preparation of a model rocket for flight is outside the scope of this book. However, we will outline the steps to complete the construction of the model rocket prior to launch. As we have already glued our motor mount in place, we will start the final assembly of our model rocket with the shock cord.

Installing the shock cord

The shock cord of a model rocket consists of a flat paper wedge and an elastic cord. The shock cord attaches the nose cone and parachute to the body tube of the rocket and provides shock absorption for the nose cone as it is jettisoned from the body tube.

For our rocket, the paper wedge has a top length of 30 mm, a bottom length of 20 mm, and a height of 80 mm:

Figure 9.28 – Paper cut-out for shock cord

To install the shock cord into our model rocket, we do the following:

1. Cut out the paper wedge (image may be found in the GitHub repository of this chapter specified in the *Technical requirements* section).
2. Glue one end of the elastic cord to section 1 of the cutout (*Figure 9.29 a*).
3. Fold section 1 onto section 2 (*Figure 9.29 b*).
4. Put glue on section 3 and fold the mount in half (*Figure 9.29 c*).
5. Allow to dry before gluing the mount inside the paper tube at the opposite end of the motor mount (*Figure 9.29 d*):

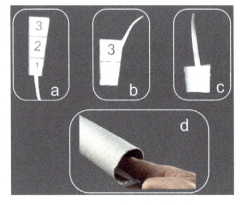

Figure 9.29 – Installing the shock cord

Ensure that the mount is placed far enough inside the paper tube so that it will not interfere with the nose cone shoulder (>20 mm).

With the shock cord installed, it is now time to glue the fin can in place on the rocket.

Installing the fin can

When we designed the fin can, we purposely gave it an internal diameter slightly larger than the outside diameter of the paper tube. This allows us to easily slide the fin can in place over the paper tube. We will secure the fin can in place with glue. White craft glue should be sufficient.

To do this, we do the following:

1. Slide the fin can over the paper tube such that the fins are pointing toward the bottom of the motor mount and away from the shock cord, allowing a space between the bottom of the fin can and motor mount (*Figure 9.30 a*).

2. Apply white glue around the paper tube in the space between the bottom of the fin can and motor mount (*Figure 9.30 a*). A light layer of glue should be sufficient.

3. Slide the fin can over the glue, lining up the bottom of the fin can with the bottom of the paper tube (*Figure 9.30 b*):

Figure 9.30 – Gluing the fin can in place

4. Let the glue dry before handling.

With the fin can in place, we can now attach the nose cone and parachute and complete the construction of our model rocket.

Finishing the construction of our model rocket

To complete our model rocket, we need to tie the nose cone and parachute to the shock cord. To do so, we do the following:

1. Thread the shock cord through the channel at the bottom of the nose cone and tie a knot (*Figure 9.31 a*).

2. Thread the nose cone through the shrouds of the parachute and tie a knot (*Figure 9.31 b*):

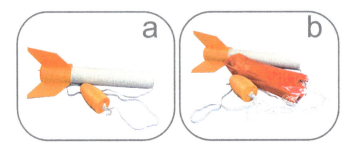

Figure 9.31 – Installing the nose cone and parachute

3. Insert the shock cord and parachute into the paper tube and install the nose cone:

Figure 9.32 – The completed model rocket

The rocket is now complete. Here is a photo of our rocket lifting off:

Figure 9.33 – Lift-off!

The preparation of the model rocket for flight is beyond the scope of this book. Many resources, such as YouTube, exist on how to prepare a model rocket for launch. The best resources are the instructions given from the manufacturers of model rocket launchpads and launch controllers, manufacturers such as Estes Industries (`https://estesrockets.com`) or Quest Aerospace (`https://www.questaerospace.com`).

As with any hobby that involves projectiles, it is imperative that we follow all local laws and safety precautions when engaging in model rocket flight.

Summary

In this chapter, we designed and built a model rocket using a standard paper tube from a paper towel roll. As measuring a paper tube can be challenging, we designed and printed a measurement tool to give us an accurate measurement of the internal diameter of the paper tube.

We used this measurement to design and print a motor mount that we installed and used to give us an accurate outside diameter of the paper tube. Using the internal and outside diameter measurements, we created a nose cone. We designed a fin can using the outside diameter measurement and glued it in place at the bottom of the paper tube. We then finished the construction of our model rocket with a shock cord and parachute.

Although this chapter was geared toward the design and construction of a model rocket, the key takeaway from this chapter should be the lessons learned in designing and printing parts around an existing shape – in this case, a simple paper tube.

In the next chapter, we will reflect on some of the ways in which 3D printing can shape the future, and we'll be designing and building a birdhouse.

Part 4: The Future

The early 2010s saw a great surge of interest in 3D printing. Many manufacturers rushed new products onto the market at an astonishing pace. The list of features began to grow and prices dropped. The 3D printer has not become one of those products that's in every home like the personal computer did a few decades ago. However, that does not mean that its impact on the modern world has been small. In this chapter, we will look at that impact and project a future where 3D printers continue to provide us with the innovation needed for an advanced society.

In this part, we cover the following chapter:

- *Chapter 10, The Future of 3D Printers and Design*

10
The Future of 3D Printing and Design

The early 2010s saw a great surge of interest in 3D printing. Many manufacturers rushed new products onto the market at an astonishing pace. The list of features began to grow, and prices dropped. Despite this, the 3D printer did not become one of those products in every home, like the personal computer did a few decades ago.

However, that does not mean that it has had little impact on the modern world. In this chapter, we will look at some of those impacts, and project a future where 3D printers continue to provide us with the innovation needed for an advanced society.

In this chapter, we will investigate the following topics:

- 3D printed homes
- The future of mass customization

In our section on 3D printed homes, we will design and build a 3D printed birdhouse.

Technical requirements

The following is required to complete the chapter:

- Any late model Windows, macOS, or Linux computer that can install OpenSCAD.

- A 3D printer with PLA – any FDM printer should work; however, the Creality Ender 3 V2 is the printer used for our example.

- A 3 mm drill tap.

- 2 M3 15 mm bolts.

The code and images for this chapter may be found here: `https://github.com/PacktPublishing/Simplifying-3D-Printing-with-OpenSCAD/tree/main/Chapter10`.

3D printed homes

Fused Deposition Modeling (FDM) printing is a simple method to conceptualize when it comes to 3D printing. It is easy to picture a print head moving in the x, y, and z directions, depositing material in layers. For the examples used in this book, the material was melted plastic that cooled into a solid soon after leaving the print head. The right combination of temperature to melt the plastic, the speed at which to extrude the melted plastic, the temperature of the print bed, and the speed at which to move the print head and bed make FDM printing possible.

It's not hard to imagine that to print larger objects, a bigger printer is required. To print an object the size of a house would require something like what we see in *Figure 10.1*:

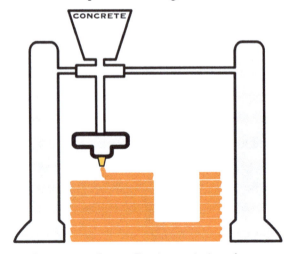

Figure 10.1 – Large 3D printer printing a house

Instead of extruding plastic, the printer in *Figure 10.1* extrudes concrete and builds the frame of a house one layer at a time. Space is left for the installation of doors and windows.

So, what exactly are 3D printed homes and how do they differ from traditionally built homes? Let's investigate these questions.

What are 3D printed homes?

Residential housing is typically built using well-known construction techniques with wood, brick, and cement. This process involves the labor of many individuals in the skilled trades' fields and may be quite time-consuming. By 3D printing the frame of a house, costs are reduced as the time needed to create the frame is reduced. The frame of a small house may be printed in a mere 24 hours.

So, to answer the question of what it is, a 3D printed house is a house where the frame is constructed layer by layer by using extruded material.

Advantages of 3D printed homes

There are many advantages to 3D printed homes, with cost being a major one. The following is a list of some of these advantages:

- As 3D printing is an additive manufacturing process, waste and, hence, costs are reduced as only the material needed for the print is used.

- The concrete used in 3D printed homes offers more strength and energy efficiency than traditional homes made from wood.

- Another advantage lies in the design options available for 3D printed homes. With 3D printing, we are not limited to right angles for our walls. A house may be designed with rounded walls and hallways, giving the house a distinctive design.

- The materials used for 3D printed homes may be sourced locally, thereby reducing transportation costs for materials, which may include recycled material. This has great benefits for more remote areas where transportation costs may be prohibitive, such as future lunar and Martian exploration camps.

Now, let's look at 3D printed homes for use in space exploration.

3D printed homes for space exploration

Creating housing for lunar and Martian expeditions is a challenging problem to solve. Even with inflatable housing, there are great transportation costs as materials need to be sent via rocket. The cost of sending a large 3D printer to remote locations on Mars, for example, can be offset by the ability to use Martian soil to 3D print the frames for buildings.

In 2015, NASA began the 3D-Printed Habitat Challenge with the goal of having teams come up with designs for 3D printed shelters on the moon and beyond. Oftentimes, NASA develops technologies that can apply to us on earth with many benefits. In *Figure 10.2*, we can see a 3D-Printed Habitat Challenge winning entry from Team SEArch+/Apis Cor:

Figure 10.2 – Entry in the NASA 3D-Printed Habitat Challenge (Credits: Team SEArch+/Apis Cor)

For our purposes and understanding, we may mimic the construction of 3D printing habitats by printing our own prototypes or tiny houses, such as dollhouses or birdhouses with a desktop 3D printer. In *Figure 10.3*, we can see a birdhouse 3D printed using a Tevo Tornado 3D printer:

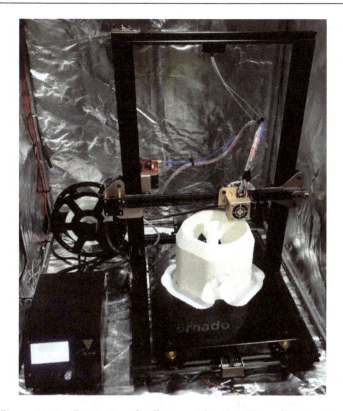

Figure 10.3 – 3D printing a birdhouse with a Tevo Tornado 3D printer

With its large build plate, the Tevo Tornado is ideal for printing mini houses, such as birdhouses. In the next section, we will turn our attention to designing and printing a birdhouse using OpenSCAD and an Ender 3 V2.

Creating a 3D printed birdhouse

Building our own 3D printed home with a desktop 3D printer may be impractical; however, experimenting with design and printing miniatures is certainly possible. Designing and printing a birdhouse allows us to experiment with design at a relatively low cost. In this section, we will design, and 3D print, a birdhouse using OpenSCAD and an Ender 3 V2. We will start with the top frame.

Generating the top frame

We will begin by generating the basic shape of the birdhouse. We will utilize the `minkowski()` and `hull()` OpenSCAD operations to create this shape. We will then hollow out this shape and add a top hook and mounting posts to the bottom.

We start by creating a new OpenSCAD file:

1. Open up OpenSCAD and create a new file. You may give the file a name such as `birdhouse.scad` by saving it before continuing.

2. Type the following code into the editor:

```
$fn=200;
screw_hole_distance = 88;


module generate_first_shape()
{
    hull()
    {
        translate([0,0,180/2])
        resize([180,180,180])
        minkowski()
        {
            cube([10,10,10], center=true);

            rotate([0,0,45])
            cube([25,25,55],center=true);

            cylinder(h=40, d1=30, d2=2, center=true);
        }
        translate([0,0,150])sphere(65);
    }
}
generate_first_shape();
```

At the heart of the `generate_first_shape()` module are the `minkowski()` and `hull()` operations. The `minkowski()` operation takes the Minkowski sum of the child nodes under it, and the `hull()` operation creates a plastic wrap-type hull around its child nodes. Two cubes, a cylinder, and a sphere are used to create the desired shape within the `minkowski()` and `hull()` operations.

3. Click on **Render**, or hit *F6* on the keyboard, to observe the first shape generated:

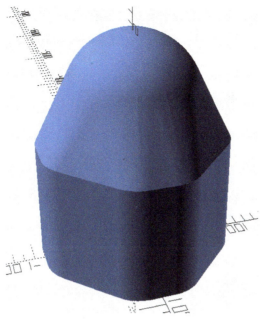

Figure 10.4 – First shape for the birdhouse design

4. Changing just one of the parameters in the `generate_first_shape()` module will alter the shape. The values used for our example have been derived from trial and error and may be changed as desired. Before we can create the module to generate the final shape, we need to add a few helper modules. Remove the line `generate_first_shape();` (the last line in the code) and replace it with the following modules:

```
module create_post_mounts()
{
    translate([screw_hole_distance,0,0])
    cylinder(h=15,d=20);

    translate([-screw_hole_distance,0,0])
    cylinder(h=15,d=20);
}
module create_hook()
{
    translate([0,0,215])
    rotate([90,0,0])
    difference()
    {
```

```
            cylinder(h=30, d=40, center=true);
            cylinder(h=30, d=30, center=true);
        }
    }
module create_screw_holes(diameter)
    {
        translate([screw_hole_distance,0,0])
        cylinder(d=diameter, h=200, center=true);

        translate([-screw_hole_distance,0,0])
        cylinder(d=diameter, h=200, center=true);
    }
```

We have used similar helper modules in the previous chapters, so we do not need to go over the code here. The modules have been given descriptive names, so their usages are self-explanatory.

5. We are now ready to add the final module that will create the top frame. Add the following code to the bottom of the editor screen:

```
module create_frame()
{
    difference()
    {
        union()
        {
            difference()
            {
                union()
                {
                    generate_first_shape();
                    create_hook();
                }
                //hollow out first shape
                translate([0,0,-2])
                scale([0.95,0.95,1])
                generate_first_shape();

                //create doorway
```

```
                    rotate([90,0,0])
                    scale([1,2,1])
                    cylinder(h=1000,  d=80,center=true);
                }
            create_post_mounts();
        }
        create_screw_holes(2.5);
    }
}
create_frame();
```

The `create_frame()` module adds a hook to the top before hollowing out the first shape and creating the door. Mounting posts are added and 2.5 mm screw holes are made on the mounting posts.

6. Click on **Render**, or hit *F6* on the keyboard, to observe that the top frame is generated.

Figure 10.5 – Top frame of the birdhouse

7. Hit *F7* on the keyboard and save the .stl file to the computer.

8. Using what we learned in *Chapter 3*, *Printing Our First Object*, print out the top frame using PLA with a generic Cura PLA profile.

> **What Is the Minkowski Sum?**
>
> Named after the mathematician, Herman Minkowski, the Minkowski sum is used in applications such as motion planning for robotics. The `minkowski()` transformation in OpenSCAD uses the second child object for addition. To really understand the `minkowski()` transformation in OpenSCAD, it is best to experiment with it.

With the top frame printed, it is time to design the bottom tray, which will give the birdhouse a bottom. For this, we will simply design a bowl-like bottom with screw holes matching the top frame.

Completing the design

For the bottom tray, we will cut a smaller cylinder out of a larger cylinder. The bottom tray will extend around the top frame, thereby providing a ledge for the birds to sit before they go inside the birdhouse.

To create the bottom tray, perform the following steps:

1. Delete the line `create_frame();` from the code and replace it with the following:

    ```
    module create_bottom_plate()
    {
        difference()
        {
            difference()
            {
                cylinder(h=20, d=215, center=true);
                cylinder(h=10, d=210);
            }
            create_screw_holes(3);
        }
    }
    create_bottom_plate();
    ```

2. Click on **Render**, or hit *F6* on the keyboard, to observe that the bottom plate is generated.

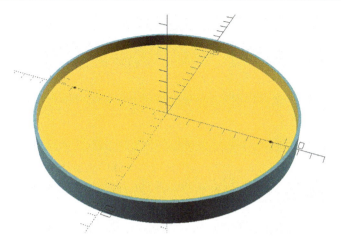

Figure 10.6 – Base plate for the birdhouse

3. Hit *F7* on the keyboard and save the `.stl` file to the computer.

4. Using what we learned in *Chapter 3*, *Printing Our First Object*, print out the bottom plate using PLA with a generic Cura PLA profile. Set **Build Plate Adhesion Type** to **None**. Alternatively, ideaMaker may be used with a texture for slicing, with **Platform Addition** set to **None**. To clear the build plate, select all objects on the plate (*Ctrl+A*) and hit the *Delete* key on the keyboard.

5. Once printed, the two parts can be joined together using M3 15 mm bolts or epoxy glue.

Figure 10.7 – Completed birdhouse

> **Generating the complete birdhouse in OpenSCAD**
>
> If desired, the complete birdhouse may be generated in OpenSCAD by adding the line `create_frame();` below the line `create_bottom_plate();` before rendering with *F6*. The birdhouse may be printed in one piece, but this is not recommended as the connection between the top frame and bottom plate would be very weak due to the horizontal layer lines.

The base plate in *Figure 10.7* was sliced with ideaMaker with the **Asian Wealth** texture applied. To get the rustic look, the birdhouse was painted with a flat black paint before a metallic wax was added for a shiny effect. An acrylic clear coat was applied to protect and provide a glossy finish.

With the birdhouse completed, it is time to look at other ways in which we may use 3D printing in the future.

The future of mass customization

The first and second industrial revolutions created much wealth and raised living standards to levels not seen before. Human-powered and water-driven machines from the first industrial revolution brought about the steam-powered machines and electrification of the second industrial revolution. With the rise of computer technology, the third industrial revolution connected computers to machines to create smarter machines.

Modern technologies such as artificial intelligence, nanotechnology (applications working with extremely small things), and 3D printing define what has been coined the fourth industrial revolution.

So, what exactly can we expect from this fourth industrial revolution and how do 3D printers fit in?

The fourth industrial revolution and 3D printing

In the fourth industrial revolution, we are introduced to the concept of the smart factory, or a factory that continues to adapt and learn. 3D printers fit well into this paradigm as they are machines that can adapt quickly to changes to a design.

As 3D printer technology continues to evolve, its place in the world of manufacturing becomes clearer. A key benefit that 3D printing brings to manufacturing is adaptability to the production line. This is especially useful when the design of a part is evolving. Changes to the design can be made between print jobs.

In addition, 3D printing excels with limited production runs due to the lack of need to create expensive and static molds. Industries such as the aerospace industry do not generally require mass production of parts.

This leads us to arguably one of the more exciting benefits of 3D printing as regards manufacturing – mass customization.

Customizing products

3D printers shine in mass customization, or the need to make parts with unique characteristics for each part made. In *Figure 10.8*, we can see a customized ring stand used to hold a championship ring. With standard mass production, each ring stand would be the same. However, in this 3D printed example, the player's name, jersey number, and stats (at the back of the stand) are unique for each ring stand printed.

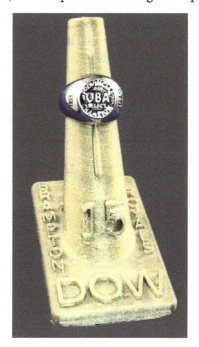

Figure 10.8 – Customized ring stand

With mass customization, products can be tailored to each customer. This is not only limited to their name – think of 3D printed footwear and how each shoe produced would fit the customer's foot perfectly.

For our purposes, it is not hard to picture taking what we learned about OpenSCAD and 3D printing and use it to create a business model based on exact customer needs and preferences.

Summary

In this chapter, we explored the use of 3D printers in the construction of homes. Although running large concrete 3D printers is beyond the scope of this book, we were able to practice our housing design skills by building a birdhouse.

We then investigated how 3D printing aligns with the fourth industrial revolution, one where smart factories constantly learn and improve. The flexibility of the 3D printing process allows for constant design improvements during production. Without the need for expensive molds, 3D printers are well suited for limited production runs suitable for sectors such as the aerospace industry.

Mass customization allows products to be tailored to each customer. Every customer, for example, may buy shoes fitted for their feet. Trophies and trophy stands may be custom-printed for each player on a team.

With this chapter, we end our journey together through learning the amazing OpenSCAD design program for 3D printing. I hope the journey has been as fulfilling for you, as the reader, as it has been for me, as the writer. It has been a true joy sharing this experience with you.

Hi!

I am Colin Dow, author of Simplifying 3D Printing with OpenSCAD. I really hope you enjoyed reading this book and found it useful for increasing your productivity and efficiency in OpenSCAD.

It would really help me (and other potential readers!) if you could leave a review on Amazon sharing your thoughts on Simplifying 3D Printing with OpenSCAD here.

Go to the link below or scan the QR code to leave your review:

`https://packt.link/r/1801813175`

Your review will help me to understand what's worked well in this book, and what could be improved upon for future editions, so it really is appreciated.

Best Wishes,

Colin Dow

Index

Symbols

T

U

W

X

Y

`Packt.com`

Subscribe to our online digital library for full access to over 7,000 books and videos, as well as industry leading tools to help you plan your personal development and advance your career. For more information, please visit our website.

Why subscribe?

- Spend less time learning and more time coding with practical eBooks and Videos from over 4,000 industry professionals

- Improve your learning with Skill Plans built especially for you

- Get a free eBook or video every month

- Fully searchable for easy access to vital information

- Copy and paste, print, and bookmark content

Did you know that Packt offers eBook versions of every book published, with PDF and ePub files available? You can upgrade to the eBook version at `packt.com` and as a print book customer, you are entitled to a discount on the eBook copy. Get in touch with us at `customercare@packtpub.com` for more details.

At `www.packt.com`, you can also read a collection of free technical articles, sign up for a range of free newsletters, and receive exclusive discounts and offers on Packt books and eBooks.

Other Books You May Enjoy

If you enjoyed this book, you may be interested in these other books by Packt:

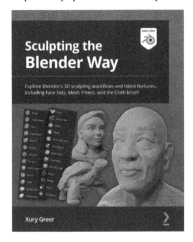

Sculpting the Blender Way

Xury Greer

ISBN: 978-1-80107-387-5

- Configure your graphics tablet for use in 3D sculpting
- Set up Blender's user interface for sculpting
- Understand the core sculpting workflows
- Get the hang of using Blender's basic sculpting brushes
- Customize brushes for more advanced workflows

Learn SOLIDWORKS 2022 - Second Edition

Tayseer Almattar

ISBN: 978-1-80107-309-7

- Understand the fundamentals of SOLIDWORKS and parametric modeling
- Create professional 2D sketches as bases for 3D models using simple and advanced modeling techniques
- Use SOLIDWORKS drawing tools to generate standard engineering drawings
- Evaluate mass properties and materials for designing parts and assemblies
- Join different parts together to form static and dynamic assemblies

Packt is searching for authors like you

If you're interested in becoming an author for Packt, please visit authors. packtpub.com and apply today. We have worked with thousands of developers and tech professionals, just like you, to help them share their insight with the global tech community. You can make a general application, apply for a specific hot topic that we are recruiting an author for, or submit your own idea.

www.ingramcontent.com/pod-product-compliance
Lightning Source LLC
Chambersburg PA
CBHW062105050326

40690CB00016B/3214